W9-BKU-564

Vincent G. Dethier

# To Know a Fly

**McGraw-Hill, Inc.**
New York St. Louis San Francisco Auckland Bogotá
Caracas Lisbon London Madrid Mexico City Milan
Montreal New Delhi San Juan Singapore
Sydney Tokyo Toronto

To Know a Fly







Library of Congress Catalog Card Number: 88-62462

ISBN 0-07-016574-2

4 5 6 7 8 9 0 DOC/DOC 9 9 8 7 6 5 4

# Foreword

IN THE WORLD TODAY, scientists are a caste, isolated from, and simultaneously hated and loved by, the community. On the one hand, scientists impress their fellow men and are admired by them—partly because their jargon sounds impressive, partly because they are seen to produce so much that *works:* airplanes, rockets, atom bombs, television, cures for dreaded diseases, hybrid corn. On the other hand, the taxpayer resents their using his money at a frightening and ever-increasing rate, and nobody can help being irked by feeling "left out"; the scientist's shop talk is as incomprehensible to most people as the hokus-pokus of the magician. Because they are misunderstood, scientists are met with suspicion. They are in danger of becoming outcasts.

Until recently, the situation was most serious for physicists and chemists. But now also for the biologist the days are over when outsiders could easily understand what he is doing—and be either bored or amused by it. Each of the many biological subsciences has amassed such a vast amount of detailed knowledge and developed such intricate techniques and such a specialized jargon that communication is possible only between the members of the guild—a handful of them for each subscience in each country. Since life processes are so much more complex than physical phenomena, this difficulty will become even more serious in the future. Biologists, therefore, are bound to be met with more suspicion than their fellow scientists.

Of course, scientists, being human, perceive this well enough. Many of them try to remedy the situation by attempting to explain what they are doing, why they think their work is worthwhile, and why they need increasing amounts of money. But many, indeed most of these attempts meet with failure, or at best with only limited success. The scientist is lucky if people react the way a farmer's wife did when she saw my graduates paint pine-cones in preparation for tests on the color vision of little sand wasps: "I suppose it's useful, or the Government would not pay you for it."

Yet, if science is to flourish, a merely tolerant community is not good enough. The community should be convinced of the need for scientists; it should be willing and indeed eager to support them. Scientists therefore should give their very best when "popularizing" their work. Those who have made the attempt have discovered not only that there is keen interest in their work, but also that they have clarified their own thinking.

Because the best scientists are as a rule too busy—driven by a strong urge to speed up the annoyingly slow process of painstaking investigation (so that they will live to see some progress), the task of popularizing science has been carried mainly by a guild of interpreters. But communication between the scientist and his fellow men should preferably be direct; people are entitled to hear the story straight from the horse's mouth.

It is fortunate, therefore, that there are men like Professor Dethier who are not merely outstanding researchers but who also have the gift of clear communication. These abilities, which Dethier possesses to an exceptional degree, ensure him a large and fascinated audience whenever he reports about his work to meetings of professional biologists; in addition, they enable him to keep open a line of communication with non-biologists. His task is in a way made easier because he happens to be interested in (and in a sense, to have fallen in love with) one of the most common animals in the world: the humble fly. Although most people will think of a fly merely as a pest to get rid of, flies are familiar to everyone and therefore make a good starting point for the non-scientist.

Among biologists, Dethier further occupies a rather unusual position. His main interest is to understand the behavior of a fly, but he has managed to avoid the almost schizophrenic split one finds in biology, which makes many investigators study either the movements of the intact animal, or the processes going on in tiny little bits of its nervous system. The students of the intact animal are often called psychologists or ethologists; those of bits of live machinery, physiologists. Dethier is neither, or rather he is both in one, and in his work he shows how research into the way animals work leads step by step from psychology into physiology, or from physiology into psychology.

This book is outstanding also in another way. It does not merely explain how one sets about the task of learning how a fly works; it also shows the research worker enjoying his work. It shows him pursuing with absorbing interest the problems that present themselves in his work. It explains why the scientist, because of his eager curiosity, at times appears childish to others. But most important, it shows that the scientist, in spite of being, perhaps, a little possessed, and at times somewhat childish, is fundamentally no different from other people with a calling in life.

*N. Tinbergen*

# To Know a Fly

"What sort of insects do you rejoice in, where *you* come from?" the Gnat inquired.

"I don't rejoice in insects at all," Alice explained, . . .

Lewis Carroll, *Through the Looking Glass*

# Chapter 1

ALTHOUGH small children have taboos against stepping on ants because such actions are said to bring on rain, there has never seemed to be a taboo against pulling off the legs or wings of flies. Most children eventually outgrow this behavior. Those who do not either come to a bad end or become biologists.

It is believed in some quarters that to become a successful modern biologist requires a college education and a substantial grant from the Federal Government. The college education not infrequently is as useful for acquiring proficiency in the game of Grantsmanship as it is for understanding biology. No self-respecting modern biologist can go to work without money for a secretary, a research associate, two laboratory assistants, permanent equipment, consumable supplies, travel, a station wagon for field collecting, photographic supplies, books, animals, animal cages, somebody to care for the animals, postage, telephone calls, reprints, and last, but by no means least, a substantial sum (called overhead) to the university to pay for all the stenographers hired to handle all the papers and money transactions that so big a grant requires. The grant, of course, must be big in the first place to allow for the overhead. Thus equipped, the biologist retires into his automated electronic laboratory. He may never see a live animal or plant. He has come a long way since the days when he pulled off the wings of flies.

It need not be, however. Anyone with a genuine love of nature, an insatiable curiosity about life, a soaring imagination, devilish ingenuity, the patience of Job, and the ability to read has the basic ingredients and most of the necessary accoutrements to become a first-class biologist. The only necessary item remaining is an experimental animal (or plant). There is much to be said for the fly.

Like taxes, the fly is always with us. As a matter of fact, there are, at the latest count, about 50,000 kinds of flies

*"And then there is overhead at 20% . . ."*

sharing "our" world. They include, to mention only a few, houseflies, fruit flies, soldier flies, snipe flies, small-headed flies, stiletto flies, blowflies, march flies, dance flies, horse flies, stable flies, black flies, tsetse flies, crane flies, humpbacked flies, bee flies, flat-footed flies, big-eyed flies, thickheaded flies, sand flies, robber flies, gadflies, dung flies, and louse flies. A motley and prolific crew!

A characteristic human reaction to flies is to eradicate them. This is deemed a meritorious action. It may be an innate tendency. I have already alluded to the propensity of children for amputating the appendages of flies. Ants do the same thing if a fly is unfortunate enough to fall into their clutches. I venture to predict, however, that the fly is in no danger of extinction. It has no sociological impediments to reproduction; its food supply is unlimited; its basic requirements, few. In some areas of the desert, for example, where there is no vegetation and little animal life except for men, goats, and camels, one dung cake may be a thriving nursery housing hundreds of flies-elect. So, if we must live together, if we cannot live in peaceful coexistence (if I dare mention the phrase), let us learn something about and from some of our fellow earthlings.

We have been educated to think of the fly as that villainous character who carries disease, a veritable Pandora whose body is an animated box of germs of a number and dreadfulness all out of proportion to its size. The tsetse fly transmits African sleeping sickness, a disease which in 1900 in one section of Africa killed 200,000 out of a population of 300,000; the mango fly transmits the disease loaiasis; the sandfly, phlebotomus fever; the blackfly, a disease which is frequently blinding. But the next time a fly lands on the edge of the dinner table to rest and to rub its feet together in anticipation, stop before you mash it into the woodwork with a fly swatter (this is one insecticide against which a fly will never develop resistance). This little beast accomplished one thing that you and I can never accomplish; he *flew* there!

One of the smaller flies weighs about seven millionths of a pound. He is equipped with two reinforced membranous wings which must serve the dual function of wing and propeller. Failure to take into account this dual function of insect wings led to the famous miscalculation proving that bees could not fly. The bee had been analyzed as a helicopter. So it is with the fly. To stay airborne and move forward, the fly must beat his wings as much as two hundred times a second. By comparison, the hummingbird beats about seventy-five times a second; the rattlesnake rattles about one hundred times a second on the hottest days. The fastest repetitive muscular contractions that you and I can produce occur at the rate of about ten times a second.

The fly has other wondrous accomplishments, too, not the least of which is being able to land on the ceiling. For years, controversy raged as to whether he managed this by executing a half-roll or an inside loop. As a matter of fact, he does neither. He flies close to the ceiling in a normal position, then reaches up and back over his head with his front feet till they touch the ceiling, whereupon he somersaults over into position. The incredible nimbleness of flies is no secret to anyone who has attempted to catch one in his cupped hand, nor is their astronomical power of reproduction to anyone who has tried to eradicate them.

When choosing an experimental animal, therefore, why settle for anything so prosaic as the laboratory rat, so giddy as the guinea pig, so phlegmatic as the frog, so reptilian as the chicken, so cousinly as the chimpanzee? Why not choose an excitingly different creature like the aardvark or the dugong? Why not choose the fly?

Chrysalis: "The whole earth is quivering,
Something mighty it is delivering,
I am being born."
Moth: "Unravel life. What are we else,
We, woven from daintiest fabrics,
But thought and soul of creation?"

Josef and Karel Čapek, *The Life of the Insects*

# Chapter 2

WITH so many kinds of flies in nature's burgeoning storehouse of life, how does one choose a proper species for study? The answer is simple. Let the species choose you. This was how our laboratory came to work with the black blowfly fifteen years ago.

As I recall, it was a steaming hot day. There was no air-conditioning in the laboratory because such nonsense was considered a luxury. The administrators had never taken into account how great an increase in efficiency would have ensued had a more suitable working environment been provided. Later we did get air-conditioning because on hot, humid days the flies in our laboratory culture died like flies. While a prostrate stenographer evoked no compassion, a cage full of dead flies constituted a powerful argument for air-conditioning in the eyes of the administration.

On this particular day, however, all the windows were open in a vain effort to catch whatever vagrant breezes were abroad. Somebody had left a liverwurst sandwich on the windowsill. It did not take it long to ripen in that heat. A gentle liver-like odor filled the laboratory and welled over to the outside. Somewhere in the wide acres beyond, an egg-burdened blowfly cruised urgently about seeking a suitable place to lay her cream-colored cylindrical eggs. In her wandering, she encountered the tantalizing odor and, unraveling the tangled skein that swirled and twisted in the lazy breeze, she came upon the liverwurst sandwich. Her busybody ovipositor sought out the crevices between the bread and the liverwurst and packed in row upon row of eggs. Twenty-four hours later we found the maggot-laden sandwich and transferred the squirming mass to a small cage.

It was both expensive and impractical to continue to raise flies on liverwurst sandwiches. And for sanitary and esthetic reasons, it was desirable to find some substitute for the fly's normal diet in nature. At first we tried ground horsemeat, but the alterations that fly maggots performed on this

COOL
COLD

material made the laboratory a perfumer's hell. On the other hand, the old saw that it's an ill wind that blows no good was never more conclusively demonstrated. For a year we had been endeavoring, unsuccessfully, to persuade the administration that our working quarters were unreasonably cramped. Suddenly we were informed that the occupants of the adjoining laboratory were being relocated on another floor and that no one was claiming the space. We hastily agreed to fill the vacuum. In time we changed the dietary regime of our flies. For the past eight years or so they have been breeding happily on a mixture of powdered milk, brewer's yeast, and agar. Aside from a moderately strong ammonia odor, the cultures are not especially objectionable.

A few of the many advantages of employing blowflies as experimental animals becomes readily apparent as soon as one begins to rear them. There is, first of all, economy. A rat costs from $1.50 to $3.00, a cat from $6 to $10, a monkey from $50 to $100. Blowflies cost nothing. It is true that human beings can be also had for nothing. They are called volunteers. The latitude of experimentation permitted with human beings is, however, quite limited. The blowfly has none of these disadvantages. Not only can the flies be procured free of charge, legally, and without a hunting license; they can besides be maintained at a fraction of the cost of what it takes to feed any other experimental animal. Thousands of them can be reared in a few square feet of space, and no caretaker is required for their maintenance.

There are two other aspects of animal experimentation that pose thorny problems for those biologists who would work with larger animals or those higher on the evolutionary

*Disposing of unsuccessful experiments is no problem.*

scale. The first is the problem of disposal. If the experiment is such that the animal is happily alive at its conclusion, one has acquired a dependent for life. There are some animals that even a zoo will not accept. A colleague of mine once carried out some experiments with a baby alligator. When all was finished and the alligator none the worse for wear, there was not a child in the neighborhood whose parents would allow him to accept the beast for a pet. My colleague did not have the heart to kill the animal, so he made room for it in the laboratory. That was nearly twenty years ago. The alligator is now about four feet long, eats untold pounds of meat a year, and has a vile disposition. The building in which he is housed is about to be torn down to make room for new construction, and nobody knows quite what to do with the alligator. My colleague thoughtfully retired several years ago.

If an experiment is a terminal one, one has a corpse on his hands. Many localities have strict ordinances governing the kinds of materials that can be placed in garbage. The only alternative is to bury the remains. If you work in a university in the heart of a great city, this is difficult. There was a time in the history of science when scientists had to sneak out in the dead of night to dig up corpses for anatomical studies. Now some frequently have to sneak out to bury their unsuccessful experiments. Both procedures are illegal. Without going further into an unpleasant subject, let us merely note that flies are no problem at all.

A much more touchy problem is that posed by various organizations devoted to the prevention of cruelty to animals. In some countries and some states research is seriously hampered by overzealous application of humane principles. But for some reason that I have never been able to understand, a line has been drawn across the evolutionary scale separating those animals that must be treated humanely and those outside the Pale. (Although just recently in a scientific journal I noticed a most unscientific article on humane ways to cook lobsters.) Insects occupy a position somewhere far down the scale. Their position is poignantly delineated by Don Marquis' cockroach, archy, who declared:

> i am going to start
> a revolution
> i saw a kitchen
> worker killing
> water bugs with poison
> hunting pretty
> little roaches
> down to death
> it set my blood to
> boiling
> i thought of all
> the massacres and slaughter
> of persecuted insects
> at the hands of cruel humans
> and i cried
> aloud to heaven
> and i knelt
> on all six legs
> and vowed a vow
> of vengeance

Could it be the insects stir no spark of pity in our breasts because they are mute creatures? Who knows their feelings, if any? Could it be that they arouse no feeling of kinship because their great eyes are motionless, pupilless, and stare out blankly on the world? Whatever the answer may be, the fact remains that he whose heart shudders at the thought of working with vertebrates has little compunction when swatting a fly.

Perfection is a will-o'-the-wisp, however, and there are certain disadvantages with working with flies. They have an uncanny knack of escaping. My laboratory is constantly host to numbers of flies that eventually congregate in the light fixtures. These vagabonds plus some German cockroaches, which originally inhabited the animal room (where colleagues keep such commonplace creatures as mice, rats, guinea pigs, and rabbits) but underwent a population explosion that carried them to the far corners of the building, tend to detract from the antiseptic appearance of the modern laboratory. Ironically, I cannot use the more efficient insecticides to control these interlopers because dust particles from the cinder block walls carry the insecticide into the very cultures I desire to protect. Our research was once set back three whole months when a well-meaning secretary authorized a local exterminator to clean up our "filthy" building. Now I control unmanageable insect populations in my laboratory with rolls of fly paper hung from points of vantage and uncoiled in dark recesses of the furniture.

Have you even tried to buy fly paper these days? Every enquiry for this item brings down upon you the pitying stare of the clerk who proceeds to educate you with toxicity data on DDT, HETP, TEPP, dieldrin, aldrin, allethrin, chlordane, heptachlor, toxaphene, parathion, malathion, lindane, diazinon, etc., etc. About the only place one can find fly paper these days is in an old country general store. I get my supply from a delightful emporium of this sort in one of the little-known coastal towns of northern Maine.

Another great disadvantage of working with flies is the small amount of food that they consume. This is especially true of the adults. They are happy with a saucer of dry sugar and a bottle of water inverted so that it stands on a piece of filter paper in a petri dish and leaks out an adequate supply over the weeks. For egg-laying, however, it is a different story. Raw liver is the most attractive medium for egg-laying. A fly that would remain egg-bound till the death when there is no suitable substrate available will disgorge hundreds of eggs in the twinkling of an eye as soon as a small piece of raw liver is provided. Simple, you say? Of course, if you wish to pay for the liver out of your household budget. If you wish to pay out of research funds, the business almost becomes unmanageable because of the small sum of money involved. It costs more to process a purchase order through proper channels than the liver itself costs. In truth, it is easier to purchase supplies for an elephant than for a fly. There is always a forlorn hope that the purchasing agent will make a typographical error which will so augment the quantity ordered that the order goes through without a hitch. It once happened to me when I ordered a carboy of glycol. In due time I received a call from the Purchasing Department informing me that my *carload* of glycol was down on the siding. It was a simple matter to sneak down one moonless night and draw off the small amount I desired.

Even the process of establishing a petty cash fund becomes so involved that you eventually class the fly as a dependent and bring the liver from home. It could be worse, however, because there are flies which are parasites in caterpillars. The adult fly seeks out leaves upon which caterpillars have been crawling and lays its eggs near caterpillar footsteps. This frequently insures that wandering caterpillars eating a leaf will also consume the eggs of the fly. The maggots then develop within the body of the caterpillar. A supply of caterpillar footsteps might command an exorbitantly high price.

One of my classmates, when we were both students studying flies, solved the problem for himself by going to a local grogshop, which was well populated with flies even in the dead of winter, and capturing his flies on the hoof. This was not without its hazards, first because the proprietor tolerated this behavior only as long as one was a paying customer, second because the flies became more elusive as one imbibed, and third because one was always suspect in the eyes of other patrons. But just as Daudet's Père Gaucher had to sacrifice his sobriety for the good of his Order, a scientist is frequently called upon to make great sacrifices for science.

i scurry around
gutters and sewers
and garbage cans
said the fly and
gather up the germs of
typhoid influenza
and pneumonia on my
feet and wings . . .

then i carry the germs
into the households of
men

. . . . . . . .
. . . it is my mission
to help rid the world
of these wicked persons
i am a vessel of right-
eousness
scattering seeds of jus-
tice
and serving the noblest
uses

don marquis, *the lives and times of archy and mehitabel*

# Chapter 3

A PROPERLY conducted experiment is a beautiful thing. It is an adventure, an expedition, a conquest. It commences with an act of faith, faith that the world is real, that our senses generally can be trusted, that effects have causes, and that we can discover meaning by reason. It continues with an observation and a question. An experiment is a scientist's way of asking nature a question. He alters a condition, observes a result, and draws a conclusion. It is no game for a disorderly mind (although the ranks of Science are replete with confused thinkers). There are many ways of going astray. The mention of two will suffice.

The most commonly committed scientific sin is the lack of proper experimental control. The scientist must be certain that the result he obtains is a consequence of the specific alteration he introduced and not of some other coincidental one. There is the case of the gentleman who had trained a flea to leap at the command "Jump!"

"Now," said the clever gentleman, "I shall do an experiment to discover where the flea's ears are located. First I shall amputate his feelers." Whereupon, the operation having been completed and the flea having recovered, the command "Jump!" was given. The flea jumped. "Ah," said the gentleman obviously pleased, "he does not hear with his antennae. I shall now amputate his forelegs." With each succeeding operation the flea leaped on command until only the hindmost legs remained. When they were removed, the flea failed to jump. "You see," concluded the gentleman triumphantly, "he hears with his hind legs."

Or there is the well-known case of the chap who wondered which component of his mixed drink caused his inevitable intoxication. He tried bourbon and water, rum and water, scotch and water, rye and water, gin and water and concluded, since every drink had water as a constant, that

*Insecticides?*

water caused his drunkenness. He then gritted his teeth and tried water alone—with negative results. When I last saw him he had concluded that the glass was the intoxicating agent, and he was about to begin another series of experiments employing paper cups.

Of course even controls can be carried to absurd extremes as in the case of the atheistic scientist who seized upon the opportunity afforded by the birth of twins to test the efficacy of religion. He had one baby baptized and kept the other as a control.

Another common fallacy is that of confusing correlation with cause and effect. This is exemplified by the case of the gentleman who was extricated from the rubble of an apartment house immediately after an earthquake. "Do you know what happened?" his rescuers inquired.

"I am not certain," replied the survivor. "I remember pulling down the window shade and it caused the whole building to collapse."

The kind of question asked of nature is a measure of a scientist's intellectual stature. Too many research workers have no questions at all to ask, but this does not deter them from doing experiments. They become enamored of a new instrument, acquire it, then ask only "What can I do with this beauty?" Others ask such questions as "How many leaves are there this year on the ivy on the zoology building?" And having counted them do not know what to do with the information. But some questions can be useful and challenging. And meaningful questions can be asked of a fly.

Between the fly and the biologist, however, there is a language barrier that makes getting direct answers to questions difficult. With a human subject it is only necessary to ask: what color is this? does that hurt? are you hungry? The human subject may, of course, lie; the fly cannot. However, to elicit information from him it is necessary to resort to all kinds of trickery and legerdemain. This means pitting one's brain against that of the fly—a risk some people are unwilling to assume. But then, experimentation is only for the adventuresome, for the dreamers, for the brave.

It is risky even at higher levels. I am reminded of the eminent professor who had designed experiments to test an ape's capacity to use tools. A banana was hung from a string just out of reach. An assortment of tools, that is, boxes to pile up, bamboo poles to fit together, etc., were provided, and the ape's ability was to be judged by his choice of method. To the chagrin of the professor, the ape chose a method that had never even occurred to that learned gentleman.

Extracting information from a fly can be equally challenging. Take the question of taste, for example. Does a fly possess a sense of taste? Is it similar to ours? How sensitive is it? What does he prefer?

The first fruitful experimental approach to this problem began less than fifty years ago with a very shrewd observation; namely, that flies (and bees and butterflies) walked about in their food and constantly stuck out their tongues. The next time you dine with a fly (and modern sanitary practice has not greatly diminished the opportunities), observe his behavior when he gavots across the top of the custard pie. His proboscis, which is normally carried retracted into his head like the landing gear of an airplane, will be lowered, and like a miniature vacuum cleaner he will suck in food. For a striking demonstration of this, mix some sugared water and food coloring and paint a sheet of paper. The first fly to find it will leave a beautiful trail of lip prints, hardly the kind suitable for lipstick ads but nonetheless instructive.

Proboscis extension has been seen thousands of times by thousands of people but few have been either struck by the sanitary aspects of the act or ingenious enough to figure out how they might put the observation to use to learn about fly behavior.

The brilliant idea conceived by the biologist who first speculated on why some insects paraded around in their food was that they tasted with their feet. In retrospect it is the simplest thing in the world to test this idea. It also makes a fine parlor trick for even the most blasé gathering.

The first step is to provide a fly with a handle since Nature failed to do so. Procure a stick about the size of a lead pencil. (A lead pencil will do nicely. So will an applicator stick, the kind that a physician employs when swabbing a throat.) Dip one end repeatedly into candle wax or paraffin until a fly-sized gob accumulates. Next anaesthetize a fly. The least messy method is to deposit him in the freezing compartment of a refrigerator for several minutes. Then, working very rapidly, place him backside down on the wax and seal his wings onto it with a hot needle.

Now for the experimental proof. Lower the fly gently over a saucer of water until his feet just touch. Chances are he is thirsty. If so, he will lower his proboscis as soon as his feet touch and will suck avidly. When thirst has been allayed, the proboscis will be retracted compactly into the

*"I've lost my appetite for pistachio mousse."*

head. This is a neat arrangement because a permanently extended proboscis might flop about uncomfortably during flight or be trod upon while walking.

Next, lower the fly into a saucer of sugared water. In a fraction of a second the proboscis is flicked out again. Put him back into water (this is the control), and the proboscis is retracted. Water, in; sugar, out. The performance continues almost indefinitely. Who can doubt that the fly can taste with his feet? The beauty of this proboscis response, as it is called, is that it is a reflex action, almost as automatic as a knee jerk. By taking advantage of its automatism, one can learn very subtle things about a fly's sense of taste.

For example, who has the more acute sense of taste, you or the fly? As the cookbooks say, take ten saucers. Fill the first with water and stir in one teaspoon of sugar. Now pour half the contents of the saucer into another which should then be filled with water. After stirring, pour half of the contents of the second saucer into a third and fill it with water. Repeat this process until you have a row of ten saucers. Now take a fly (having made certain that he is not thirsty) and lower him gently into the most dilute mixture. Then try him in the next and so on up the series until his proboscis is lowered. This is the weakest sugar solution that he can taste.

Now test yourself. If you are the sort of person who does not mind kissing his dog, you can use the same saucers as the fly. Otherwise make up a fresh series. You will be surprised, perhaps chagrined, to discover that the fly is unbelievably more sensitive than you. In fact, a starving fly is ten million times more sensitive.

You console yourself with the thought that he may be

less versatile, less of a gourmet, than you. Well, this too can be tested. Try him on other sugars; there are any number of sugars: cane sugar, beet sugar, malt sugar, milk sugar, grape sugar. Each is chemically different; each has for you a different sweetness. It is only necessary to determine for each the most dilute solution that will cause the fly to lower his proboscis. Then when the sugars are listed in order of decreasing effectiveness, it turns out that the order is the same for you and the fly: grape sugar, cane sugar, malt sugar, milk sugar, beet sugar. In one respect the fly is less gullible; he is not fooled by saccharine or any other artificial sweeteners.

But, you may argue, I can distinguish many other kinds of tastes. This is only partly correct. You can distinguish many kinds of flavors, but to assist you in this you recruit your nose. Flavor is a mixture of tastes, odors, and textures. With taste alone you are pretty much restricted to sweet, salt, sour, and bitter.

The old adage that one can catch more flies with honey than with vinegar has a sound basis in physiology. Leaving aside for the moment the fact that flies react differently to different odors, the truth remains that flies accept materials that taste sweet to us and reject those that taste salt, sour, or bitter to us. This fact, too, can be demonstrated with the proboscis response, but the only way for a fly to say "No" is to retract his proboscis, and it can be retracted only if it is first extended. Accordingly, one prepares several saucers of sugared water. A pinch of salt is added to one, two pinches to another, three pinches to a third, and so on. As before, the fly is lowered gently into the saucer with the least salt. He responds, as expected, by extending his proboscis. He is then allowed to taste the next dish, and the next, and the next. At one of these dishes he will stubbornly refuse to extend his proboscis. Since this dish contains the same amount of sugar as the rest, one must conclude that it is the salt that is being rejected. The test can be repeated with vinegar, lemon juice, or quinine water. It can even be tried with aspirin, whiskey, bicarbonate of soda, tobacco juice—anything that will dissolve in water. If you wish to be really sophisticated, you can test the relative sensitivity of his legs and mouth by standing him in one solution and allowing his proboscis to come down into a different one. A friend of mine who once wished to study the stomach of the fly and to color it so it could be seen more easily under the microscope hit upon the idea of standing a fly in sugar but arranging for its mouth to come down in dye. As a result the fly's insides were stained beautifully. This is one example of a physiological way to coat a pill.

"And what does it live on?"
"Weak tea with cream in it."
"Supposing it couldn't find any?" she suggested.
"Then it would die, of course."
"But that must happen very often," Alice remarked thoughtfully.
"It always happens," said the Gnat.

Lewis Carroll, *Through the Looking Glass*

# Chapter 4

WHETHER he realizes it or not, the average person *is* interested in the feeding habits of insects. After all, every one of our troubles with insects arises from their feeding habits and nothing else. Mosquitoes, fleas, bed bugs, horseflies, no-see-ums interest us only because they bore into our hides and steal our blood. Corn borers, Colorado potato beetles, aphids, scales, cabbage worms excite our interest only because they eat our food before we can get to it. Clothes moths and carpet beetles try to turn all our fabrics into Belgian lace. Termites feed on our houses and libraries. "And you will not wonder," wrote Wee-Wee, 43rd Neotenic King of the 8,429th Dynasty of the Bellicose Termites to Professor William Morton Wheeler of Harvard, "at my knowledge of some of the peculiarities of your society when I tell you that in my youth I belonged to a colony that devoured and digested a well-selected library belonging to a learned missionary after he had himself succumbed to the appetite of one of the fiercest tribes of the Kamerun." We are not particularly interested in their flying, their sleeping, or their courtship—although some charming anecdotes abound on the latter score.

There is, for example, *Mormoniella*. *Mormoniella* is a minute wasp that lives during its youth as a parasite within a developing fly (pupa). More than one wasp may grow within a single fly. Since the pupa represents a limited supply of food, the eventual size of the wasp depends upon how many share the same fly. If there are too many, they are miniatures. Now, *Mormoniella* engages in elaborate courtship. The valiant swain courts his betrothed by playing pattycake with his antennae and hers. After a while, in an ecstacy of antennal swordplay, their union is consummated. In the case of a midget male and a normal-sized female, the little fellow, because of his short stature, is unable to court and consummate simultaneously. He must do one or the other, and since one is a *sine qua non* for the other, he never succeeds.

A biologist, who was shamelessly eavesdropping on all of these activities, had the puckish idea of allowing two midget males to court a female simultaneously. The long and short of it was that while one little fellow was courting, the other was consummating the union. To this day the biologist has been unable to decide whether he had witnessed an example of cooperation or of competition.

But let us return to matters of taste. The entire crew of insect gluttons and gourmets have essentially the same kind of taste organs—hairs. The fly is a marvelously hirsute beast. One usually thinks of hair as something to be combed, brushed, and eternally cut or to be removed from animals and turned into coats. In short, one tends to think of hair as purely ornamental. Animals, if they think about it at all, think solely in terms of its insulating properties in wintertime.

With the fly it is different. With his skeleton on the outside like a mediaeval suit of armor, his only way of sensing changes in the outside world is to have sensitive projections sticking out all over him—especially if he is to sense events before it is too late to do anything about them. One is reminded of the various projections and sensing elements on satellites.

It is not altogether surprising, therefore, that flies taste by means of hairs. It would be even less surprising to you if you were to look at your own (preferably somebody elses unless you are a contortionist) tongue through a miscroscope. Your own organs of taste are equipped with minute hairs.

It requires only a reasonably strong magnifying glass to satisfy yourself that the fly tastes with certain of his hairs.

With the addition of a steady hand, one can satisfy himself that only the very tip of the hair is sensitive. Some further requirements for this experiment to succeed are abstinence from coffee and other stimulants on the day of the experiment and a bountiful supply of invective.

First you take a sewing needle. Only a new one works because, as I found out later, it retains a thin film of machine oil. Next take a mixture of rubbing alcohol and sugar and place a drop on the needle by deftly flicking the needle in and out of the solution. A supple wrist aids greatly. Now, under the microscope or magnifying glass, in the nearby light of a strong lamp, coax the drop from the needle to the base of one of the larger hairs clothing the end of the proboscis. If you succeed in this, slowly roll the drop up the shaft of the hair. While this is going on, the heat of the lamp begins to evaporate the alcohol in the drop with three beneficial results: the amount of alcohol (unacceptable to the fly) is decreased; the concentration and hence sweetness of the sugar is increased; the drop gets smaller and smaller. If your timing has been correct and your hand steady, you will have succeeded in rolling the drop to the summit of the hair. Now expel your breath—not in the direction of the fly! You will have observed that only when the drop of sugar reached the tip did the fly extend its proboscis. And were you to mix a bit of salt with the drop you would discover that the same hair tastes both salt and sugar.

There is a very old law of biology which states essentially that a sense organ sends only one kind of message to the brain regardless of the kind of stimulus it receives. Thus, a punch in the eye sends the same message as a flash of light.

*The sexes are most easily distinguished by facial expressions.*

The brain interprets all messages from the nerve connecting it to the eye as having been caused by light, hence with a punch one "sees" stars. Even if the sense organ is no longer there, only the stump of its nerve, the brain interprets any activity in the nerve as having come from the sense organ. A person who has lost a leg may complain of a pain in his toe. To him the sensation from his phantom limb is overwhelmingly real. The brain interprets irritation in the stump of the nerves which formerly ran to the toe as still arising in the toe.

From one sense organ there should be only one sensation, light from the eye, taste from the tongue, sound from the ear. When there are different qualities from within these sensations which can be distinguished, it can only mean that there are different kinds of cells making up the sense organ, each normally sensitive to a special kind of stimulus. Thus, the hair of the fly must really be at least a dual sense organ. Without even being able to see its cells one can predict that it must have at least two. But the same hair actually responds not alone to salt and sugar but also to water and bending. If a fly is thirsty, water on the hair causes a proboscis response; if a fly is ravenously hungry, gentle bending of the hair causes a response. How many nerve cells does the hair have? With ingenuity and no equipment more expensive than a magnifying glass, one can answer this question without ever setting eyes on the cells. The trick is to take advantage of a phenomenon called adaptation.

Adaptation is nothing more or less than getting used to a condition. Even children are aware of this, as witness the little girl about to go into a movie theatre with her younger sister whose eyes are firmly closed. "Why," asked the usher in the lobby, "has she got her eyes closed?"

"Well, you see," answered the older of the two, "when we get in the dark, she opens her eyes and finds a seat."

Sometimes adaptation occurs in the sense organ itself as in this case; sometimes it occurs at different levels in the brain; sometimes both are combined. After you are dressed in the morning you are no longer conscious of the feel of clothing against your body; after a dull lecture or sermon has progressed *ad nauseam* you no longer hear the voice. As soon as it stops, however, you are yanked back to the present by its absence. It is amazing how the silence at the conclusion of a university lecture awakens the students whose parents are paying so expensively for this opportunity to sleep.

Certain natural phenomena, such as adaptation, can be useful as tools to understanding. Take, for example, a thirsty, hungry fly. With a fine needle gently twang one hair. Out comes the proboscis. Continued titillation elicits repeated extension—up to a point. Has the fly stopped because it is fatigued? Shift your attention to a neighboring hair. As soon as this is bent the proboscis extends with its usual alacrity. Obviously the nerve cell in the hair has become adapted. Before it recovers, place a drop of water on the hair. Immediately, the hair, which was made insensitive to bending, responds to water. Again continue applying water until adaptation ensues. The hair is now insensitive to bending and to water. At this juncture, apply sugar. Once again the proboscis extends. The only reasonable interpretation of this simple experiment is that the hair is equipped with one nerve cell sensitive to bending, one to water, and one to sugar.

Finally, since salt elicits retraction rather than extension, there must be a separate cell for salt.

Here, then, in the hair the fly possesses a sense organ whose simple appearance and small size belies its complexity and information-carrying capacity. At this point one cannot help but wonder with awe and humility at the order, beauty, and complexity of the universe.

a louse i
used to know
told me that
millionaires and
bums tasted
about alike
to him

don marquis, *the lives and times of archy and mehitabel*

# Chapter 5

A CERTAIN executive constantly faced with the task of making decisions broke down so completely one day that he was banished for therapeutic purposes to a potato farm as a common laborer. The farmer, however, recognizing a man of superior intelligence, soon promoted the ailing executive to the grading shed where his job consisted of sorting potatoes by age, perfection, size, and so forth. At the end of the first day, the foreman checked on the new worker to see how he was doing. He found the poor devil sitting before an Everest of potatoes, one specimen in each hand, muttering over and over to himself, "Decisions. Decisions. Always decisions."

Even without its psychotic aspects, the act of choosing is undoubtedly one of the least understood phenomena in the world. What we really mean by choice, how it operates, how one can ever prove that such a thing as choice even exists are questions that plague people in all walks of life. The biologist wonders if lower animals can choose, and if so, how far down the animal family tree the ability goes. If a fly can do it, does this mean that a brain capable of reasoning is not required, or does it mean that a fly can reason? It is often said that choice is a matter of taste, so I decided to apply the old saw literally. My opportunity arrived one June in the personage of a prospective graduate student who expressed a desire to work under my aegis. At one and the same time I would assign him a research problem on choice by the fly and employ the results as a yardstick in making a decision as to whether or not I should choose him as a student.

This was probably as sound a procedure as any. Although colleges and universities maintain beautifully equipped admissions offices and avail themselves of batteries of intelligence and aptitude tests to assist in the selection of students most likely to succeed, the intangible something that distinguishes a potential creative research worker from

*"There's one advantage in our inability to make decisions—
we never make the wrong ones."*

all other students can be truly detected only by giving a candidate a research problem to incubate and hatch. A potential artist can be detected only by the picture he paints, a writer by the composition he writes, a hen by the eggs she lays, a research worker by the research he does. One doesn't devise a test to see if a hen can lay an egg.

Accordingly, when my student presented himself, I looked neither at his scoring in the innumerable national tests he had had to take nor at his undergraduate record. Instead I said, "If you can come back in the fall with a solution to this problem, I'll take you on." As an afterthought I added, "I'll bet you a bottle of beer you fail."

The problem was this: how much does a fly eat a day and what factors regulate his intake? When I returned to the campus in September the student was camping on the doorstep of my laboratory with the answer.

Mr. X (that wasn't his name) had brought with him a one-quart preserving jar with a two-piece lid (the kind with the removable inner section). He had replaced the inner part of the lid with a piece of copper screening into which he had cut two small holes. He then picked up in the laboratory two pieces of glass tubing about ten inches long. He drew out the end of each tube in a gas flame until the opening was about the thickness of pencil lead. Each tube was then bent into the shape of a J with the small orifice on the turned-up end. He then filled one tube with water and the other with sugar, marked the top level of fluid in each, stood the two in the preserving jar with their long ends sticking up through the holes in the screen cover, and by sleight of hand managed to insert twenty flies before tightening the screw-top lid. The next day he removed the flies, carefully took each tube, inserted a hypodermic syringe into the small opening on the turned-up end, and pumped in water until the fluid level climbed to the mark he had made the previous day. Then he merely read from the syringe how much fluid had been required to refill each tube. In each case this was the amount of the particular fluid consumed by twenty flies in twenty-four hours. It was all so absurdly simple I began to wonder if scientific affluence had dulled my acuity.

When both tubes contained the same fluid, the flies drank equal volumes from each. When one contained sugar and the other water, the flies took more sugar. By this simple inexpensive technique, we could learn still more about the fly's sense of taste. For example, by testing more and more dilute sugar solutions against water we found a concentration of sugar below which the fly could not discriminate between water and sugar. By testing more and more concentrated solutions, we found that the stronger the solution was the more the fly took. Strangely enough, the fly apparently had no mechanism for regulating its calories or the concentration of its body fluids.

Psychologists who have studied rats in similar situations have discovered that the rat tapers off as the sugar concentration is increased. Most humans—children and certain neurotic adults excluded—will control the amount of sweets they ingest and will find some foods just too sweet to take. For the fly there is no limit.

Since the fly exhibits feeding preferences, it is even possible with this simple technique to check our earlier experiments showing that some sugars are "sweeter" than others.

One need only place cane sugar in one tube and an identical concentration of glucose in the other to learn that cane sugar is the more acceptable of the two—as it is to us.

Flies and some people are very much alike in another respect; both prefer what tastes good to what is nutritionally best. Although flies ignore saccharine and all other artificial sweeteners, they are gluttonous over a rare sugar known as fucose. Given a choice of glucose or fucose, flies gorge themselves on fucose and slowly starve to death even though there is a more than adequate supply of glucose in the neighboring tube a mere one inch away. Here is an example par excellence of eating being a matter of taste. Presumably they die happy.

With imagination, there is no end to the wealth of information to be gained from the choice experiment. Some questions, however, demand new techniques. To learn more about the actual method of choice one must know the number of visits a fly makes to each tube and the duration of each drink. Originally we had established an indoor fly-watching society for this purpose. Its members included students who would do anything for a good grade, except study, and laboratory technicians who would do practically anything for a raise.

Only rarely, however, is the scientific hireling any more reliable than the biblical hireling with his sheep—especially if the job is killingly monotonous. Watching flies in a bottle is not the most stimulating pastime in the world. It is astonishing how much of a fly's life is spent doing nothing. A fly sits, rubs its front legs together, its front legs over its head, its front legs and its middle legs, its back legs over its wings, its back legs together, and on and on. It crawls, it flies, it drinks, it makes fly specks. But mostly it just sits. Such a wanton waste of time is devastating to the morale of one who himself can think of so many better ways to use time.

If only automation would replace fly-watchers. It now being June again, I called in Mr. X. Since he now enjoyed all the rights and privileges of a graduate student, such as being an overworked and underpaid sheep in the academic wilderness, I had to think of some new bribe. For the promise of a case of beer (plus the one bottle that had been a forgotten promise) he engaged to spend the summer dreaming up a way to record fly drinks automatically. This time I thought I had him, but I had underestimated the thirst of a graduate student.

In the fall he presented me with a solution to the problem. There was the same old preserving jar, the same two tubes—with additions. Down the entire length of each tube and completely circumscribing the small opening of the drinking end was a layer of conducting paint—that paint which conducts electricity and is used in the printed circuits of some cheap radios. Near the top of each tube a piece of wire was soldered to the paint. Inside each tube, extending from end to end and out the top, was another length of wire. The two wires from each tube then circled sloppily around in the air and finally disappeared into an amplifier that was attached to a recording instrument that wrote on graph paper. At the moment the pens were tracing beautiful symmetrical wavy lines.

My student had taken advantage of a condition in all laboratories that drives everyone mad who works with delicate electronic equipment. This curse is the sixty-cycle electricity that is always present as leakage from light circuits, motors, refrigerators, air-conditioning units, elevators, etc. The fly-drinking equipment faithfully picked up this electrical noise and recorded it. When, however, a fly stood on the painted tip of the tube and lowered his proboscis for a drink, he short-circuited the sixty-cycle with no ill-effects to himself. Instead of an undulating line, the pens then traced a straight line. A straight line represented a drink!

With the success of this device assured, I paid my debt in beer and disbanded the fly-watching society.

"Well, I should like to be a *little* larger, Sir, if you wouldn't mind," said Alice. "Three inches is such a wretched height to be."

"It is a very good height indeed!" said the Caterpillar angrily . . .

Lewis Carroll, *Alice in Wonderland*

# Chapter 6

SIZE is a state of mind. Many people who are genuinely interested in such small animals as insects are intimidated by their minuteness. Others just cannot comprehend what small size really means. I remember once seeing the inside of a space rocket and being told by my engineer guide how every inch of space had been utilized and how beautifully minute all the working parts were. The thought that crossed my mind was that there was still space for a goodly number of cockroaches to stow away and that each cockroach had more working parts and was more versatile than the whole rocket.

People seem to adopt a negative attitude toward small size (even in people) and yet ingenuity overcomes practically all size problems. There was one biologist, for example, who was interested in the flying mechanism of insects. He was able to place a fruit fly, an insect about ⅛ inch long, in a chamber where he could measure its respiration during flight and at the same time equip it with the following recording instruments: platinum electrodes to record the nerve impulses to its wing muscles, a glass needle connected to a phonograph pick-up to record the mechanical movement of the muscles, and a thermister to measure changes in body temperature. At the same time the rate and amplitude of wing strokes were being recorded by other devices.

More recently there was published an account of experimentation with a cricket into whose brain had been implanted electrodes so that the brain could be stimulated with weak electrical currents. Depending on where the electrodes were placed, various kinds of behavior could be elicited. The cricket could be made to rub its legs, sing, walk, or jump. Jumping was the least successful response, and the ultimate explanation shows how sympathetic one must be with his experimental animal.

The cricket was attached to a stationary holder. He stood on a little cork ball. When he walked it was as though

*When he takes the bit in his mouth, the best thing to do is to give him his head.*

he were on a treadmill; when he tried to jump, however, ballistic recoil sent the ball off into space and the cricket could not "land." Even though he was fixed, the normal conclusion to jumping is landing. Once the experimenter realized this he provided a parabolic mirror against which the cork ball was projected when the cricket jumped. The mirror deflected the ball back to its starting point, and the cricket "landed."

Although the fly is smaller than a cricket, size is really no handicap. It is no trick at all, for example, to do fly surgery. On second thought, perhaps it is all trickery. You can judge for yourself. We had observed that the sensitivity of a fly to sugar depended upon whether or not he had recently fed. We became curious about the states of hunger and satiation in the fly. These are questions that also attract the attention of physicians, psychiatrists, and vertebrate physiologists. At another level they are of interest to most of mankind.

We had observed that the sensitivity of a fly to sugar increased nearly ten million-fold from the time he was fed until one hundred hours after feeding. In other words, as he became hungrier he would respond to more and more dilute sugar so his sensitivity was a good measure of his state of hunger. One hundred hours, however, is a long time to wait, so we devised a method of making flies hungrier quicker. This is very simple. One merely flies the fly to exhaustion. At the end of an hour or two he has used all his food reserves as fuel and is ravenously hungry.

But how does one fly a fly? Stop and think of the conditions experienced by a flying fly: his feet hang in midair; he sees the ground moving by; he feels wind blowing past his face. To duplicate these conditions we attached with wax a short thread to the back of a fly. This in turn was fastened to a stick that would revolve around a fixed point like a merry-go-round. The thread was long enough so that the fly could stand on a phonograph turntable. When the turntable was started, the fly's feet were yanked out from under him and in his vision the ground began to move, so he began to fly. This caused him to be his own merry-go-round. With the wind generated by his own movement and the moving pattern of the turntable beneath him, he flew until no fuel was left.

An even simpler way to fly a fly, but one which requires more courage, is that of walking him as one would walk a dog. As we walked our flies up and down the corridors of the laboratory, people stared and shied away. The fly was only a blur at the end of a thread that stuck out horizontally in apparent defiance of gravity. The real trouble came in attempting to turn the fly around at the end of the corridor for the return trip.

At any rate, we readily produced hungry flies. But what is hunger? The possibility occurred to us that the difference between a hungry fly and a fed fly was the amount of sugar in its blood. The first way to tackle this question was to measure the blood-sugar levels of a fly in various stages of hunger. We employed precisely the same methods that hospital laboratories employ for measuring sugar levels in the blood of people. We merely miniaturized everything to accommodate the very small amount of blood that could be expressed from a fly. The best way to appreciate the smallness of the blood volume is to see for yourself by pulling one leg off a fly and squeezing him.

As we had more or less expected, a hungry fly had less sugar in his blood than a fed fly. Then we conducted an experiment to test the hypothesis that some causal relationship existed between blood-sugar level and hunger. We injected sugar into a hungry fly in much the same way that a hospital patient receives intravenous glucose. To our astonishment the fly remained hungry. We tried more sugar, and more, and more. After we had produced candied but still ravenously hungry flies, we were forced to conclude that blood-sugar levels did not regulate hunger or satiation.

As a clincher we made what is known as parabiotic twins. Two hungry flies were surgically joined together back to back so that they shared a common blood supply. One fly was then fed and both subsequently tested for sensitivity to sugar to determine how hungry they were. The fed fly was satiated; his unfed partner, hungry.

In human beings one of the manifestations of hunger is movement of the stomach—the so-called hunger pangs. A manifestation of satiation is a feeling of fullness. With these facts in mind we attempted to find out whether the presence or absence of food in the alimentary canal of the fly altered its behavior towards food.

The alimentary canal or gut of the fly is rather different from ours. Immediately in back of the neck, where it enters the thorax, the gut branches. One branch is a blind sac, the crop; the other, the mid-gut, leading into the small and then the large intestine. It is possible to open up a fly and observe where and how fast food travels. It can be seen that food passes primarily into the crop where it is stored. After feeding, it is transferred from time to time from the crop to the main part of the gut.

By making a small incision, much as a surgeon does to remove an appendix, we were able to reach into the fly and either tie off the crop or mid-gut before or after feeding, or even remove these parts entirely. By experiments of this sort, we showed that neither the presence nor absence of food in these regions or even the regions themselves had anything to do with hunger and satiation.

Just to make sure, we loaded the gut with food, artificially. It is not possible to feed the fly with a stomach tube, but fortunately for us as well as for the fly, there is an opening at both ends. We decided to give the fly a super enema. The presence of some sharp S-curves in the hindmost regions of the gut complicates matters, but with luck, perseverance, and an unlimited supply of flies we successfully violated enough flies to demonstrate to our satisfaction that rectal feeding of a hungry fly failed to satiate him.

There is not much fly left. We had ruled out the blood and all of the alimentary canal except a small portion in the neck region. Perhaps the mechanism controlling hunger exists there.

It would have been nice to have injected food directly into the neck region. Unfortunately, the esophagus here does not possess the properties of a self-sealing tank as does the rest of the gut. But there is another way to test whether or not a portion of the body is involved in some activity; that is, to interrupt its nervous communication with the brain. The surgical procedure for accomplishing this is a bit tricky.

For an operating table we used a block of paraffin with two small shallow depressions in it. One depression corresponded to all of the fly's body except the head. There was a separate depression for the head, but it was purposely placed too far in front of the first depression. A fly was now made to lie prone in the big depression and was held there with a strip of modeling clay pressed across his body cummerbund fashion. The head was now pulled forward as far as the stretch of the neck would allow until it could be fitted into the front depression. The head was held there by two pins crossed behind it. As a result of this procedure the fly's neck lay stretched across the bridge of paraffin separating the two depressions. It was then in position for the operation. We called this the guillotine position.

An advantage of working with flies which I failed to mention earlier is the low cost of surgical instruments. One needs only two cheap forceps honed to fine tips, a supply of insect pins—the kind used to pin specimens in boxes—and some old razor blades. The pins are prepared for use by blunting their points on the table top or by dropping them point first on the floor. As a result of this cavalier treatment, some of them are bound to acquire properly angled hooked tips. The razor blades are broken haphazardly into fragments, some of which inevitably emerge with the proper shape to serve as microscalpels.

The first move is to incise the skin on the back of the neck. The tension to which the skin has been subjected by stretching exposes the underlying organs. It is astonishing how much is crammed into this narrow space. First there are oblique and longitudinal muscles. These must not be dam-

*Careful bending of pins produces precision surgical instruments.*

aged. Since the fly is an invertebrate, hence lacking a backbone, the only thing that really holds the head on is these muscles. Gently tease them aside. Next to emerge in view are the two great breathing tubes supplying oxygen to the head. They adhere to the esophagus. Since they are the sole route of oxygen to the head, they also must be maintained intact. Gently break the adhesions and tease the air tubes to one side. Next is the heart, or actually its extension, a long tube supplying blood to the head. It lies on the esophagus. Lying under it is the nerve from the brain to all parts of the gut. It is too small and transparent to be seen in the living fly, but it is there. Reach gently under the heart and grasp the empty space where the nerve ought to be. If you indeed have the nerve, gentle pulling causes movement in some of the attached parts. Now snap it with a vicious jerk, push all the organs back into place, release the head, which now snaps back into position closing the incision, and, keeping a tight hold on the fly, remove his restraints. This last move is crucial. Many a fly, beautifully operated on with great labor, has flown off the operating table and out the window. *Sic transit gloria Muscae.*

The results of this operation on a hungry fly were spectacular. Such a fly began to eat in the normal fashion, but did it stop? Never. It ate and ate and ate. It grew larger and larger. Its abdomen became so stretched that all the organs were flattened against the sides. It became so big and round and transparent that it could almost be used as a miniature hand lens. It was so round its feet no longer reached the ground and so heavy it could not launch itself into the air. Even though the back pressure from a near bursting crop

EXPLODING FLY
KILLS U of P PROF
XXX
SUGAR
XXX
SU
VERITAS
XXX
SUGAR
bill clark

was terrific, the fly continued in its attempts to eat. It reminded me of a woman who had been admitted to our hospital, a woman whose height was four feet, ten inches and whose weight approached four hundred pounds. Her major complaint was inability to move.

We had pinpointed the mechanism controlling hunger in the fly. As long as there was some food in the region of the neck, the nervous system sent messages to the brain stopping further eating. When no messages were sent, eating resumed. This was hunger. By interrupting the pathway to the brain, we had made the fly chronically hungry.

i can show you love
and hate
and the future
dreaming side by side
in a cell
in the little cells where
matter is so fine it
merges
into spirit
you ask me where i
have been
but you had better
ask me where i am
and what i have been
drinking
exclamation point

don marquis, *the lives and times of archy and mehitabel*

# Chapter 7

ALTHOUGH all the animals that live on land today are descended from cells that once lived in the water, they have never quite been able to put their aquatic ancestry behind them. They have never quite been able to get along without water. Some are more enslaved than others as, for example, the toad which must each spring return to the ponds and swamps to lay her eggs. Most of the other animals have shaken free of this compulsion, but even though they no longer bathe the outsides of their bodies in water, they cheat, in a manner of speaking, by periodically sluicing water through themselves. This is known as drinking. Some members of the human race have striven mightily to free themselves entirely of the necessity of imbibing water by substituting other fluids, but this endeavor has met with limited success.

Animals are at best imperfect vessels, and this imperfection extends to the inability to retain water for any long period of time. It is lost at an alarming rate through a variety of exits. The leakiest part of an animal is most obvious to anyone who is concerned with sanitary problems in the young of the human species. Less obvious is the fact that the opposite end also loses water. The only advantage of which I am aware of losing water from the head end is that one can polish spectacles by first breathing on them.

Another route for water loss has been seized upon by certain segments of the business world who try to impress the public with the social desirability of blocking this route. It used to be said that horses sweat, men perspired, and ladies glowed, but modern advertising has stripped from this concept whatever dignity it may have had. In any event, the loss of water through the skin is by no means minimal in certain animals.

Different species of animals show individuality in their patterns of sweating surface. In the days when horses were essential components of urban life and not museum pieces

*The fly is ubiquitous.*

or elements of sporting events, nearly everyone knew that they sweat profusely over most of their bodies. Most people realize that dogs lose water by panting rather than by sweating. Few people know what cats do. Guy Gilpatrick, in one of his Glencannon stories, wrote of a bet between the chief engineer and the mate of the tramp steamer, Inchcliff Castle, as to if and how cats perspired. In a true scientific spirit, the bet was to be settled by experiment.

Glencannon had maintained that cats perspired through the soles of their feet so to test the hypothesis a cat was to be heated and then persuaded to walk across a piece of blotting paper. For reasons most plausibly explained by Gilpatrick, the experiment fell somewhat short of success.

Practically nobody understands the problem of water loss in the fly. Flies do not perspire, yet they lose water so readily that they are living virtually in a desert all their lives. A situation that you and I might find unbearably humid can be one in which a fly dries up completely overnight. This is one of the hazards of small size. A small creature has much more surface area per unit of body volume than a large creature, and excessive surface area usually means excessive water evaporation—and also heat loss. This is one reason, for example, why small children out skating or skiing with their parents invariably become so cold so quickly.

The obvious way for a fly to forestall drying up is to drink copiously, but how does it know when it is drying up? In other words, how does a fly know when it is thirsty? A person may become thirsty not only by losing water but by gaining salt. One of the subtle ways to get a person to drink more beer is to provide him with free pretzels—well salted. We

*"Everything I drink has the same effect."*

tried a comparable trick with the fly; that is, mixed salt with its drink in order to see if its drinking would increase. No change occurred, however.

Still thinking about the situation in human beings, we decided to see if we could induce thirst by making the fly lose water. There are two ways to accomplish this end; one is to place the fly in a closed container with some very dry material that will absorb water. Best for this purpose is calcium chloride. A fly subjected to this treatment will lose more than twenty per cent of his body weight in twenty-four hours. That is equivalent to a 150-pound man losing thirty pounds overnight. A more heroic means is to bleed the fly. I have mentioned flies' blood before. It is scanty and colorless, not red. Very few animals, as a matter of fact, possess red blood. In insects it may be colorless, yellow, or green; in lobsters, crabs, and Bostonians, it is blue. Regardless of its color, no special technique is required to remove it from small animals. As any child will tell you, you pull off a leg and squeeze.

Following either drying or bleeding, a fly becomes thirsty. Again, the situation reflects that in man. One has only to recall that in every novel or cinema the wounded hero who has suffered a loss of blood forgets being a hero long enough to be physiological and cry for water. Replacing blood should allay the symptoms of thirst. In the fly it certainly does. But, and here the fly differs from man, replacement of blood with any fluid achieves the same end. We have injected into thirsty flies blood, water, concentrated salt, mineral oil, and even lighter fluid. Each of these satisfies the fly's thirst; that is, he no longer drinks water. The lighter fluid eventually kills him, but at least he does not die thirsty.

The difference between the fly and ourselves is one of osmotic pressure. Osmotic pressure is a term that a majority of college students find beyond their comprehension. For that matter, many textbooks on Introductory Biology do such a miserable job of explaining it that one suspects the authors themselves of having difficulty. Perhaps it is best explained by analogy. When the mythical Dutch boy poked his finger in the hole in the dike, he had to push out as hard with his finger as the water was pushing in, in order to prevent a flood. Now an animal, from one point of view, is just a bag of water with many living dikes—membranes separating various bodies of fluid. The walls of the blood vessels separate the blood on the inside from the fluid bathing the outside; the walls of the stomach and intestine separate the fluids on the inside from those on the outside; the wall of every cell in the body separates the cell sap on the inside from the various watery fluids surrounding the cell.

Most living membranes are leaky dikes. They are riddled with holes so small as to be invisible even under the microscope but large enough, nontheless, to permit the passage of some molecules. If the amount of salt in the fluids on either side of a membrane is different, and the holes in the membrane permit the passage of water but not salt, water will tend to accumulate on the side with the greatest amount of salt. The pressure that one would have to apply against the holes in the membrane to prevent this build-up of water is termed osmotic pressure.

In most animals, man included, there must be just the proper amount of water in the various compartments of the body. Changes of salt concentration will cause water to pass where it should not, causing the animal distress and very probably death. This is one reason why fluids injected into the blood of a human being must have precisely the correct concentration of salt or sugar. For some reason not yet clearly understood, the fly is not bothered by these problems. I hasten to add, however, that there are differences even among flies. The tsetse fly, which feeds exclusively on blood, is almost immediately killed by a drink of water. It seems that there is after all a modicum of truth in the assertion of topers that water is toxic.

When we injected a variety of fluids into our thirsty fly and satisfied his thirst, we showed beyond reasonable doubt that the fluid pressure (not the osmotic pressure or salt concentration) in the body cavity was the means of preventing the fly from acting thirsty. When this pressure dropped, the fly drank water.

Water is a rare commodity in some circumstances. One of the really beautiful sights in the tropics is clouds of butterflies swirling over puddles or camp areas in a jungle clearing. Someone in the desert may be pestered beyond reason by flies in his eyes, nose, and mouth, drawn there for the moisture. In these and many other cases there must be a mechanism in the wee body that tells the creature when to drink and when not to drink.

personally
no matter what
the season
i can always find
a ragout
where i may drop in
for a warm bath
and a bite to eat

don marquis, *the lives and times of archy and mehitabel*

# Chapter 8

ONE of the tragic but classical examples of the occasional superiority of the wisdom of the body over the wisdom of medicine is the case of the boy who craved salt. This child would consume fistfuls of salt; one of the first words he learned to speak was "salt"; the most cherished object in the house was the salt shaker. This unnatural diet so alarmed his parents that they brought him to a hospital for observation. Here he was placed on a standard hospital diet, and here he shortly died. Autopsy revealed that two small glands in the region of the kidneys were diseased, hence unable to function properly. Under the circumstances the body required abnormally large quantities of salt. The boy had kept himself alive by satisfying an intense craving for this particular food.

Particular cravings are not the product of disease alone. It is a standard joke that the husband of a pregnant woman may be routed out of bed in the wee matutinal hours to fetch some exotic morsel, pickled pawpaw, stewed guava, or a pistachio nut sundae. On a less exotic plane there is the craving for salt that one experiences after excessive sweating in the heat or the taste for sugar that develops after exercise on an empty stomach. In each case a particular need is met by satisfying a craving for a particular food.

Precisely how the mechanism of specific hunger operates is a mystery. It is not a phenomenon confined to the higher animals, and therefore may conceivably work in a simple manner. Flies exhibit a specific hunger. The first inkling of its existence derived from observations in nature by people who were interested, for reasons best known to themselves, in collecting flies. These collectors soon discovered that male flies were usually found flitting around flowers while female flies congregated more often around dead or decaying material. A number of conclusions may be drawn from this observation, one being that males, no matter what they are, are irresponsible playboys. Were it not for the fact that a few

*In the spring a young fly's fancy lightly turns to thoughts of protein.*

males occasionally frequent decaying materials for social reasons, one might almost conclude that they are at best ornamental adjuncts to society.

Another conclusion that might be drawn is that males prefer to imbibe nectar, which is predominantly sugar, while females prefer meat, despite the nursery rhyme describing what little girls are made of. Do not allow the fact of the female's diet being in a state of decay trouble you. If you enjoy Camembert or Limburger cheese, or hung venison, jugged hare, or similar delicacies, you are in no position to criticize. In some parts of the world decayed food is not for gourmets alone; it is standard fare. Certain African tribes would not be caught dead eating fresh fish or meat although you and I would be caught quite dead if we ate flesh that was not fresh. In the old Gold Coast "stink fish" was synonymous with "excellent" when applied to food. In a Congo meat market I once saw some large chunky black objects offered for sale. An up-wind approach was impossible so I edged cautiously down-wind to inquire as to their nature. They were venerable chunks of elephant meat, blackened with age and held together either by an olfactory strength or by all of the resident germs holding hands.

In any event, there are times when the female fly has a hunger for protein. It can be demonstrated quite scientifically, although far less exotically, by the same method I have already described for measuring food intake. For the present demonstration, one of the J-shaped pipettes contains protein and the other, sugar. The choice of a suitable protein for testing may be left up to the experimenter. Although a thawed liver mousse is greatly preferred by the flies, I do

*"My dear, I owe it entirely to my strict protein diet."*

not recommend it. Liver in a warm laboratory has a tendency to acquire an exquisitely vile odor. Unfortunately this is true of most protein.

If you have the fortitude to measure a fly's daily consumption of sugar and of protein over a three-week period, you will discover that the males always prefer sugar. The females, on the other hand, prefer protein to sugar during the first six or eight days of life; thereafter, they, too, prefer sugar. A series of simple dissections will reveal that it is during the first week of life that a female is building her eggs. The development of a fly's eggs cannot proceed to completion without protein any more than a hen's egg can be packaged in a shell by a hen that lacks calcium in her diet.

After the eggs are fully developed, the fly switches her preferences from protein to sugar where it remains until death unless the first batch of eggs is laid and another batch started. We now come to the basic question: what causes the change in feeding preference? By means of the surgical techniques I have described earlier, we have come closer to an answer. However, in order to make this part of the story understandable I must digress long enough to mention hormones.

Hormones in man, for example, are special chemicals produced by the thyroid, adrenal, pituitary, gonads, and other glands whence they are metered into the blood stream to influence body growth, development, and activity. It is a judicious balance of these secretions that makes possible the deep voice and heavy beard of men and the beauty we so admire in women. It is the malfunction of the glands that produces giants, dwarfs, and various pathetic abnormalities.

No less in the fly are there hormones controlling growth and reproduction. Eggs will not develop in the absence of hormones from a gland in the neck; this gland will not produce its hormones in the absence of a gland embedded in the brain. And as is true of so many biological systems, there are reciprocal effects whereby the sex organs of the fly exert influence on the glands in the neck and brain.

To one interested in protein preferences all of these interrelationships present a knotty problem. How does one go about discovering the mechanism causing the fly to select protein at certain times in her life? Since the preference is coincidentally related to egg development, one might hazard a guess (the scientific jargon is "propose a hypothesis") that the preference is caused by the developing eggs. The first step toward testing this guess is to remove the ovaries. The surgical techniques developed earlier make this absurdly simple (with a little practice). Ovaries can be removed by an operation superficially resembling an appendectomy. The result? No change in protein preference. So one makes another guess. Since egg development is related to hormone production, perhaps the preference is under hormonal control. The gland in the neck can be removed by the same technique employed in cutting the nerve from the gut. The gland in the brain can be removed by an operation that involves cutting a minute hole in the back of the head, reaching more or less blindly into the proper region of the brain, and cutting out that piece in which the gland is known to lie. For comparative purposes it is interesting to note that the entire fly brain is smaller than the size of the cut usually made in human brain surgery.

Any one of these operations halts the development of eggs but fails to abolish the protein preference. The specific hunger does not seem to be under direct hormonal control. The next guess is that the level of proteins in the circulating blood affects either the organs of taste or some point in the brain where their nerves connect with the feeding mechanism. And this guess, too, will eventually be tested, and should it prove false, the next and the next, for this is how our understanding progresses.

Even short of the goal, we seem closer to an understanding of the specific hunger in the fly than we are to comprehending how the specific hungers of man work. And we know already that there is a time when the female fly prefers protein, which cannot nourish her own body, to sugar, which is an adequate food for her but useless for her eggs. Here is an example of survival of the individual being subordinated to survival of the species. In some quarters it would be hailed as maternal instinct, and by so naming it we would be no nearer an understanding of what it is.

So Mr. Daddy Long-legs
    And Mr. Floppy Fly
Sat down in silence by the sea,
    And gazed upon the sky.
They said, "This is a dreadful thing!
The world has all gone wrong,
Since one has legs too short by half,
    The other much too long!"

Edward Lear, *Nonsense Songs, Stories, Botany and Alphabets*

# Chapter 9

THE charwoman who comes in to clean our laboratory at night is neither brighter nor duller than the average American, and she has a rather confused idea as to what goes on during the day. Since we who work there are obviously scientists, we must, in her mind, have something to do with rockets and space, and since at the same time we hold the title "Doctor," we devote ourselves to curing the bodily ills of mankind. The fact that we choose to experiment on flies does not bother the dear lady one whit. With blind faith she trusts our judgment implicity in all things. As I said, she is a person of average intelligence. But recently she has come to 'look upon us with a rueful suspicion that we may be in league with the devil. Her disillusionment came about as a result of something the flies told me about her working habits.

To appreciate this situation you must know something about the events leading up to the revelation. By this time we knew a great deal about the fly, his sense of taste, his preferences, his hungers, his thirst, his satiation. Meditating one day on these facts I got to thinking that being hungry served no useful purpose unless hunger also stimulated one to search for food. A hungry fly should be more restless than a happily fed fly. The technical problem boiled down to this: how does one measure restlessness in a fly? There is always the method of direct observation, of course; but this is incredibly tedious and time consuming. Insects in general seem to spend enormous lengths of time waiting. I've often wondered what goes on in the microscopic brain during these periods. When not waiting, insects wander aimlessly. And yet, strangely enough, this aimless wandering is always an indirect way to something important. It is by aimless wandering that an insect efficiently covers vast quarters of space to find food, or a mate, or a place to lay eggs. People, by contrast, generally move directly from one place to another, usually to do something unimportant.

An efficient way to measure restlessness in a fly is to persuade it to operate some counting device, to operate a treadmill, or to use a pedometer. It should be feasible to do this since insects are notoriously strong. Fleas, grasshoppers, and springtails (those little snow fleas that spring mighty distances, usually into maple sirup buckets hanging on trees, by tucking their tails under their stomachs and slapping them against the ground) can jump distances equal to twenty or more times their length. The latest broadjump record for man, about twenty-seven feet, is only about four and one half times his own height. Cicada-killer wasps are able to fly with a captured cicada weighing easily five times their own weight.

This appearance of strength is an illusion, however, made possible in part by the small weight of the insect with reference to gravity. The absolute strength of an insect such as a fly is so small that the operation of most counting devices is a physical impossibility. Something very delicate had to be designed for our tests. The most delicate mechanical devices are balances, so we set about to design a balance that a fly could operate, one that would tell us how often he walked, jumped, or flew.

After many trials and many failures we finally succeeded in constructing a cylindrical cage of sheer nylon about one and one half inches long, mounted on a thin balsa wood frame, and balanced at the center on jeweled bearings from a watch. A needle-thin strip of brass ran the length of the floor of the cage. From it at each end hung a short piece of wire, wire so fine that even a Swiss watch-maker would not be able to drill a hole through it. Beneath each wire, mounted on the stand that balanced the cage, was a minute pot of mercury. Fortunately for us, flies do not run around a cage from top to bottom; instead they pace from end to end like proper zoo animals. Accordingly, the cage teeters like a seesaw. Everytime it tilted one of the two wires dipped into its mercury pot closing an electrical circuit which activated a pen writing on a moving strip of paper. To give an aura of scholarship to the proceedings, someone tacked on the wall above the machine the familiar lines from the Rubáiyát:

> The Moving Finger writes; and, having writ,
> Moves on: nor all thy Piety nor Wit
> Shall lure it back to cancel half a Line,
> Nor all thy Tears wash out a Word of it.

The very first day's record revealed something we had known for a long time but also something unexpected and puzzling. Flies, like the majority of insects, are creatures of light. Light exerts such a profound effect on insects that moths have been flying into candles for centuries for the edification of poets and moralists, insects in a room congregate at windows to the dismay of housewives who strive to maintain a sterile house, and honeybees placed in a test tube with its open end away from light belie the intelligence ascribed to them by romanticists and never escape because of the irresistible attraction of light. Thus, it was hardly a surprise to find on examining the record of the seesaw each morning that it gave a faithful indication of the hour of sunrise and of sunset. When the light of the rising sun reached into the laboratory each morning, it stirred the laggard flies to activity. For the entire day each fly paced back

*"I must have the brain of a moth!"*

and forth in his individual cage until at sunset the light retreated from the laboratory back somewhere to the world outside. Then activity ceased, and the recording instrument silently traced a smooth line.

This smooth line should have continued until the following morning. Instead, every night at eight o'clock the flies resumed activity for approximately thirty minutes. One night the burst occurred at nine. Then the answer to the mystery dawned on me. When the charwoman came in to clean, she turned on all of the lights and the flies became active until she left. Usually she came at eight; one night she had been an hour late. Confronted with the chiding accusation that she had been tardy one night, the puzzled woman readily admitted the fact, but because she knew that nobody had been in the building her curiosity demanded to know how I had known. I committed the very grave error of telling her the truth, or at least part of it. I told her that the flies had given her away. Perhaps I should have enlarged on this point, but she did not give me the opportunity. Crossing herself and muttering something about Beelzebub, Prince of flies, she retreated in haste and alarm.

The fact that flies are active in the daytime and inactive during the night did not particularly suit my purposes. After all, hunger does not automatically disappear at sundown and reappear in the morning (although the reverse is true in some human beings who suffer from a curious malady known as nocturnal eating). The overwhelming influence of light had to be removed. The simplest way to do this was to keep the flies either in constant darkness or in constant light. When we tried the first alternative, the flies still showed great activity during the day and rest at night *even though they were in total darkness!* The interesting fact is that flies which have been reared in total darkness for many generations and have never seen light still showed a diurnal rhythm. The rhythm, however, bears no relation to day and night. Some flies began to exercise at eight A.M., some at eight P.M., some at noon, some at midnight, etc., etc. But if they were all subjected to a flash of light at the same time in their dark world, they all synchronized their rhythms. They set their activity clocks. In nature it is the sunrise that naturally sets the clocks of all flies.

Flies do indeed have internal clocks. If you stop to think about if for a moment, you will realize that most animals have clocks. There is something inside of them that measures time. The clocks are not necessarily daily clocks; some are set to run on lunar schedules, some on seasonal schedules. The lunar schedules have captured the imagination of most people, perhaps because such interesting rhythms are based upon them. There is, for example, the Palolo worm that rises from the depths of the ocean at just the right phase of the moon and indulges in the most eye-popping orgy of reproduction. Then there is the reproductive cycle of humans, a twenty-eight day cycle, a lunar cycle in past evolution.

These cycles are controlled by some sort of biological clock. There is some mechanism within the animal that tells it when it is time to act. But fascinating as these cycles are, they interfere with, rather than help, our attempts to ascertain whether or not a hungry fly is a restless fly. Putting the flies in constant darkness failed. Since they are such slaves to light, constant light might keep them at such a high level of activity that differences due to internal rhythms would be swamped. And so it turned out. Under constant light we had more or less constant activity, a constant baseline which we could observe before and after feeding.

And then we found indeed that a hearty meal stopped activity. As a control experiment to show that the inactivity was truly a result of feeding and not merely a consequence of the added weight we burdened some hungry flies with little weights equaling a full meal. These only seemed to make the flies more active.

The behavior of the fly fits a widespread and age-old pattern. A hungry animal moves. An individual moves, a tribe moves, a race moves. They move until they find food or drop from starvation. But everyone knows this, you might argue. It is only common sense. True enough, but by showing that it occurs in a fly we establish that there is something biologically basic about the phenomenon and that the mechanism for doing it, whatever it is, is present in the small unintellectual body of a fly. Furthermore, until the fact was demonstrated experimentally it might have been an illusion.

I am reminded of an article that once appeared in a parody of a well-known scientific journal. The article had to do with the apparent increase of flies at tea-time. It went somewhat as follows: Either there are more flies at tea-time or there are not. If there are not, then there are no grounds for scientific investigation. If there are, it means either that the flies are flying faster and enter the visual field more often or are flying slower and so remain in the visual field longer. In either case we are dealing with an optical illusion.

We not only lost a charwoman, we destroyed an illusion.

There was an old person of Skye,
Who waltz'd with a Bluebottle fly;
They buzz'd a sweet tune, to the light of the moon,
And entranced all the people of Skye.

Edward Lear, *More Nonsense, Pictures, Rhymes, Botany, Etc.*

# Chapter 10

MOLIÈRE, in the first act of *Bourgeois Gentilhomme,* wrote. "Without the dance a man would not know what to do." This statement may reflect the attitude of certain specialized members of the human race but hardly applies to mankind as a species. In the insect world, however, the statement is literally true. Molière could not have known this, since the only insects with which cultivated men of his era were on intimate terms of acquaintance were fleas and lice, neither of which dance but each of which could, on occasion, incite their hosts to do so. Other insects less domesticated than the flea and the louse do dance; indeed, to dance is to exist. For some, dancing is an integral part of courtship without which procreation is impossible. Certain species of flies, appropriately known as Dance Flies, long ago anticipated one genre of night club dancing; they dance clothed in nothing but small balloons which they have produced by blowing out bubbles of a white viscus substance. Dances performed by other members of the fly clan are less spectacular; nevertheless, they play a prominent role in courtship.

For the honeybee, dancing serves a different though equally important purpose. It is an absolute prerequisite for feeding. Here is an insect that must dance for its supper. Without the dance it would not know where its supper was.

The honeybee has received better notices than other insects so most of you are undoubtedly already familiar with its performance. A foraging honeybee, having stumbled afield upon a bountiful supply of nectar, behaves rather differently than the average modern man in analogous circumstances. Instead of being secretive about the whole thing, the bee returns to the hive to spread the glad tidings. Lacking a language, she accomplishes this by performing one of a repertoire of dances. The dance that has received rave notices in the press is the "Waggle Dance," a solo dance performed without accompaniment. The premiere danseuse pirouettes

*Their internal clocks are amazingly accurate.*

figures of eight and, in traversing the waist of the eight, wiggles her abdomen in a manner marvelous to behold.

Although the esthetic qualities of this dance leave something to be desired, its excitatory effects on the honeybee audience is electric. Spectator bees mob the dancer in a frenzy of excitement until finally they rush out of the hive to the sweet cornucopia in the field. And they go exactly to the correct location! They are able to do this because the dance that is performed in the hive serves as a code giving the direction and distance of the food from the hive. When the dance, which is performed on a vertical surface in the hive, is oriented so that the bee in traversing the waist of the "eight" is crawling upwards, it means that bees leaving the hive must fly directly toward the sun to find food. When the dancing bee is crawling downward in traversing the waist of the "eight" (that is, the "eight" is lying on its other side), the foraging bees must fly away from the sun to locate the food. If the waist of the "eight" is at some particular angle with respect to the floor of the hive, 45 degrees for example, it means that the searching bee must fly a course 45 degrees with respect to the sun in order to find food.

Superimposed on this code is another which tells the distance that must be flown. The more frenzied the dance, the closer the food.

The privilege of witnessing a honeybee dance is reserved to other honeybees and to those few persons who happen to have a spare hive lying around the house. For the person whose only bees are in his bonnet or who is just naturally timid, but still wishes to see an insect dance, I recommend the fly.

*"I think she's trying to tell me something."*

According to the dictionary, a dance is a rhythmic and patterned succession of movements. There is no restriction on what is to be moved (consider the interesting varieties of dances performed by various human cultures and imitated in the name of art by our own), but most often the feet are involved. One would imagine from a purely statistical point of view that the more feet a creature possessed the more complicated the dance it could perform. An insect with its six feet should be able to generate gyrations quite beyond the capability of any human being. On the other hand, the movements concerned could involve the path traced by the animal in its entirety. This is the case with the fly. Its dance is not particularly inspiring, but then neither are many of the dances performed by the human species. In the fly the dance at least performs a useful function.

Under ordinary circumstances, a fly when not flying or standing still walks in a series of short straight lines as though it lived in a cubistic world. As it goes along, it is not only feeling the ground upon which it walks, it is tasting it. Whenever one of its six feet steps into something tasteful, the fly stops, turns in the direction of the foot in question and lowers its proboscis to the spot. This is the picture, seen, as it were, in slow motion. Actually the fly is such a neurotic, jerky creature that act succeeds act in rapid succession. The proboscis is constantly being lowered as though the fly were a miniature pile driver.

If a fly were to encounter a minute drop of sugar and some unkind biologist removed the drop as soon as the fly had had the briefest of tastes, the poor beast would begin to dance. It reminds one of a person dancing in anger or frustration. The dance lacks perfection of geometry, but it is undeniably rhythmic and patterned. The fly gyrates in ever widening circles, clockwise and counterclockwise. It spins very much like a small clockwork toy. It appears for all the world as though it were searching for the lost drop of sugar. One really cannot say that it is searching for there is no way of knowing. In the sense that a rocket with a homing device searches for an aircraft, the fly searches, but just as the rocket has the device built into it, so does the fly. One might say that the fly searches because it has no alternative. This automatic response to the cessation of a satisfactory taste is just the kind of locomotion that would most effectively lead the fly back to a drop from where it might have inadvertently strayed.

But this dance for all its crudity as compared with the Waggle Dance of the honeybee does code certain information. For example, the stronger the drop of sugar the more frenzied the dance and the longer its duration. Furthermore, the hungrier the fly, the more active and sustained the dance. One has only to clock a few dances by flies of various degrees of hunger or a few dances set off by sugar solutions of different strengths in order to be able to look at a dance and interpret it. Other flies apparently cannot decode the dance. They do not go off searching for sugar, but after all there is no place to go since the fly is not a social insect.

The only thing that flies do by way of being social is to change the food upon which they are feeding so that it is more acceptable to other flies. This again is a fact that you can test for yourself. Take, for example, a saucer containing a mixture of corn meal mush and sugar. Cover one half of the dish with aluminum foil. Set the saucer some place where flies can readily find it, in a barn, for example. After the flies have had a go at it, shoo them away, rotate the dish (as a control against the possibility of their learning position), tear off the foil, and observe which side of the corn meal is visited by the greater number of flies. You will discover that the side on which the flies had previously fed is the more attractive.

The fly dance seems designed solely to serve the individual in that it is an efficient search pattern. However, since all social behavior is inevitably constructed from elements of individual behavior, it is possible that some simple built-in dance, a primitive searching behavior, was the evolutionary forerunner of the honeybee's Waggle Dance. Perhaps if we could understand more about these primitive dances we could come closer to comprehending the dance "language" of the honeybee.

Ours is the path between two
blades of grass

Josef and Karel Čapek, *The Life of the Insects*

# Chapter 11

WE ONCE had a very ingenious lad working in our laboratory. It appeared that he had taught some ants to read. The evidence was there on the top of the laboratory bench for everyone to see. A long line of minute cinnamon-colored ants which originated from some nook or cranny in the wall (cinder block for economic reasons) hustled along the table top toward a jar of jam until they came to a neat little sign on top of a tooth pick stuck in a gob of modeling clay. When the hurrying ants came up to the sign, they made a sweeping turn to the right until they encountered a second little sign at which they turned back to their original line of march and the jam jar. The first sign read "Detour →"; the second sign, "← Detour."

Fortunately for mankind, ants cannot read. The explanation of the detouring ants is really very simple. Our lad had observed the line of ants marching across the table top and directly up the side of the jar. Knowing that some ants follow a chemical trail which they themselves deposit, he decided to trick the rest of the laboratory personnel. Without moving the jar from its place, he rotated it 180 degrees on its axis. Now, when the ants coming across the table encountered the jar, the trail stopped. Naturally they started walking around the base of this obstacle and in so doing encountered the trail going up the opposite side. Adapting to the new condition, they had soon laid a new trail around the base of the jar. Our lad now made one more change; he rotated the jar 180 degrees around the point on the base where the trail began ascending. In other words, the ants no longer had to go around to the far side of the jar in its new position in order to find the trail to the summit, but having laid the detouring trail, they continued to follow it. It was a simple matter now for our prankster to erect a couple of detour signs at the appropriate sites.

A trail is a series of signs or a continuous sign. It may be blazes on trees, cairns in the desert, a path in the pasture.

*"Life would be much less complicated if I worked for Professor Doe."*

It may be route signs at highway intersections, the line of bread crumbs that Hänsel and Gretel dropped in the forest, or the thread that Theseus unrolled to find his way back out of the labyrinth at Crete. It may be the odor of the master's footsteps that his faithful dog follows, or it may just be the sight of a person in front who is following another person who is following another, and so on, in a long queue.

Insects resort to all of these devices and then some. The ordinary tent caterpillar that disfigures wild cherry trees in the very early spring with its soiled silken tents deposits silken trails as it goes, playing a continual Sir Walter Raleigh to itself. The silken threads serve as trails, and at the same time they provide footing. They lead out from the tent to the outermost twigs where hang the succulent leaves.

Certain other caterpillars walk along head to tail like elephants in a circus procession. The celebrated naturalist Fabre recounts the tale of a group of these processionary caterpillars that marched head to tail around the rim of a circular container without breaking rank until exhaustion and starvation decimated them.

Other insects, like the ants already mentioned, deposit chemical trails. As far as we know, the fly does not deposit a trail; nevertheless, he is able to follow an artificial trail. One day in the laboratory I astounded my students by showing them a fly so well educated that he could write my name in fine Spencerian script (an accomplishment beyond the capability of many of my students). To accomplish this I first removed the wings from a hungry fly so that he would not stray from the sheet of paper on which I planned to turn him loose. Before class I traced my name on the paper with a fine camel's hair brush dipped in dilute sugar. As soon as the trace dried it became invisible. When the time came for the demonstration, I attached to the tail of the fly a minute wick dipped in ink and then released him in the vicinity of the invisible trail. After wandering a bit, his feet encountered the sugar deposited on the paper. The rest is history.

How does a fly follow a trail? Close observation reveals that he tends to walk in more or less straight lines until something diverts him. If one of his feet encounters sugar he stops, turns toward the side stimulated, and begins to feed. When he has eaten what is within the reach of his mouth, he steps forward. As long as both feet are equally stimulated, he eats his way forward in a straight line. If he gets slightly off course so that the foot on one side is off the trail, he automatically turns toward the foot that is still being stimulated. In other words, by constantly striving to keep right and left feet equally stimulated, he is able to follow the twists and turns of a trail.

It is amusing, and instructive, to draw a Y-shaped trail and turn a fly loose at the stem of the Y. What will he do at the intersection? Slavishly he eats his way along the stem, eats a bit from one arm, and then, since his left and right front feet are equally stimulated, one by the left arm and one by the right, he eats a bit from the other arm. Thus he nibbles back and forth from one arm to the next until he has progressed so far that the distance between the two arms exceeds the distance spanned by his two front feet. At this juncture he continues up whichever arm he happens to be eating at the time of losing contact.

Flying insects are in some respects at a disadvantage. By and large they cannot follow trails because a permanent trail cannot be laid down in the shifting breezes. This is probably one reason why honeybees have evolved a communications dance. On the other hand, odors in the air are useful for navigation. Think of the times you have sought out something burning in the house or a dead house mouse simply by following your nose. Recently we demonstrated the efficacy of the method of orientation in one of the suburbs of Philadelphia in which a bubble gum factory was located. On one warm night when a gentle breeze was blowing, we deposited a student who was unfamiliar with the area within a mile of the factory and asked him to go to it. He started off sniffing the gum-scented breeze. He tramped up one street, down another. Every time the odor decreased in intensity he turned back. Eventually he came to the factory, by which time he was thoroughly sick of the odor of bubble gum.

Insects employ air-borne odors to find food and mates. Fabre wrote admiringly of the immense distance over which the males of certain species of moths are able to detect and orient to the female perfume. Later workers have calculated that the odor is so dilute, about one molecule per cubic foot of air, that orientation based on a comparison of odor intensities is impossible. Some popular writers striving for sensationalism at the expense of common sense and accuracy have invoked all sorts of radio waves in an effort to explain the paradox.

A single explanation, more in accord with the facts, is available. It has been known that many insects, unless they are on some important business, fly a haphazard course with respect to wind direction. In short, they tend to ignore the wind. If in their wandering, however, they pick up an interesting scent, they cease their patternless flying and begin to fly up-wind, but only so long as the scent reaches them. If the scent stops, usually meaning that they are off trail, they again wander haphazardly until it is again encountered. In this way, without orienting to the scent itself but merely employing it as a device to keep them headed into the wind, they are able to reach a goal.

Other flying insects employ visual landmarks. A wasp that has just constructed a new nest makes a short familiarization flight in the vicinity of the nest before going out to search for prey with which to provision the nest. In one nine-second flight a wasp can familiarize herself well enough with her surroundings to be able to find her nest again. Anyone who is intrepid enough to play with wasps can demonstrate this ability to his satisfaction by moving the nest a few feet or, with greater safety, by relocating such landmarks as stones, twigs, bushes, etc.

At greater distances, flying insects navigate by means of the sun. By being able to fly with the sun in a fixed position, over the left shoulder for example, and by having a time sense that enables them to compensate for the change of the sun's position with the advancing day, they can orient with greater accuracy than the average urbanite turned loose in the country.

But some insects, notably the honeybee and the ant, are able to orient on cloudy days when the sun is not clearly visible. This accomplishment so confounded early biologists that they proposed all sorts of theories even including the one that insects could see through the clouds during the day. Theories are fine as tools to understanding but are not in themselves contributions to truth. Any clever scientist can sit down, marshall the facts at hand, and bounce out of his arm chair with a theory. The scientist who is great is the one who proposes a theory and then attempts to prove or disprove it rather than the one who proposes a theory and then goes off grinning to greener pastures leaving the onerous job of proof or disproof to others.

As far as navigation without a sun is concerned, a famous Austrian biologist was able to demonstrate that honeybees could detect differences in the pattern of polarized light in the different quadrants of the sky and hence orient correctly to their hive or to a known source of food.

All of these navigational accomplishments on the part of insects raise the interesting question of how the insects come by these abilities. This in turn brings up the tricky question of learning, which it is now time to consider.

as a representative
of the insect world
i have often wondered
on what man bases his claims
to superiority
everything he knows he has had
to learn whereas we insects are born
knowing everything we need to know

don marquis, *the lives and times of archy and mehitabel*

# Chapter 12

NOBODY really knows what learning is. This ignorance is displayed most blatantly in the field of education. Not only is there little understanding of the phenomenon of learning, there is in addition a colossal confusion between learning and teaching. Untold millions are spent annually on teaching fads, frills, gimmicks, devices ostensibly designed to make a student learn. And all too often, "learn" simply means storing facts in the memory like unused furniture in a dusty attic. One of the most discouraging features about teaching is the widespread conviction among students that the process of learning is something bequeathed by the teacher, by a book, or by TV to the student. It is a startling revelation to many that learning is an active endogenous process, that learning is done by the individual by himself. In one view, learning is the capacity to profit (*sensu latu*) by experience. It is a process by which some event happening to the organism changes the nervous system so that a response to a future event is affected by the change. Nobody understands the nature of these changes nor in which part or parts of the nervous system they occur.

It is obvious from these introductory considerations that there may be different levels of learning. Most people will agree that their dog or cat or canary learns. Many people agree that their goldfish learns. Few people have pets lower on the animal scale than these, so the question of how far down the animal scale learning goes worries only biologists. Even among this group there is not unanimity of opinion as to where learning stops. The majority are content to draw the line somewhere around the worm level.

One of the simplest types of learning is termed habituation. It is defined as ignoring a continued stimulus that is not harmful. It is seen in the purest form in churches and college lecture halls. The stimulus in these cases is the human voice.

*"Inability to learn never handicapped me a bit at the University."*

Another type of learning is called conditioning. The classical example of conditioning was that performed on dogs by the Russian scientist Pavlov. Briefly, he rang a bell, then presented a dog with food, then measured the amount of saliva the dog produced at the sight, smell, and taste of the food. After repeated trials, it was only necessary to ring the bell in order to induce the dog to salivate. Similar scenes are enacted every day in institutions still possessing dinner bells.

There is a slightly different kind of conditioning in which an animal, usually a rat, is trained to associate a bell or light with food. When the bell rings or the light glows, the rat presses a lever that releases food. Pinned on the walls of many laboratories engaged in studies of this sort is a famous cartoon in which one rat in a cage is remarking to his roommate, "Boy, have I got this guy trained. Every time I press the lever he gives me a pellet of food!" More often than not it is indeed a question of who is conditioning whom.

Higher forms of learning need not concern us here because most people have a nodding acquaintance with them even if their interest is merely vicarious.

But the fly, I regret to state, is colossally stupid. Worms learn, snails learn, cockroaches learn. We have been unable in fifteen years of work to detect any unequivocal learning ability in flies. It is possible, of course, that we are using the wrong techniques. You cannot teach a whale to fly no matter how hard you try. Furthermore, a technique that works with one animal may not work with another. A common fallacy is to make a rat learn a maze and then make a student learn a similar but appropriately larger maze. The rat turns in a

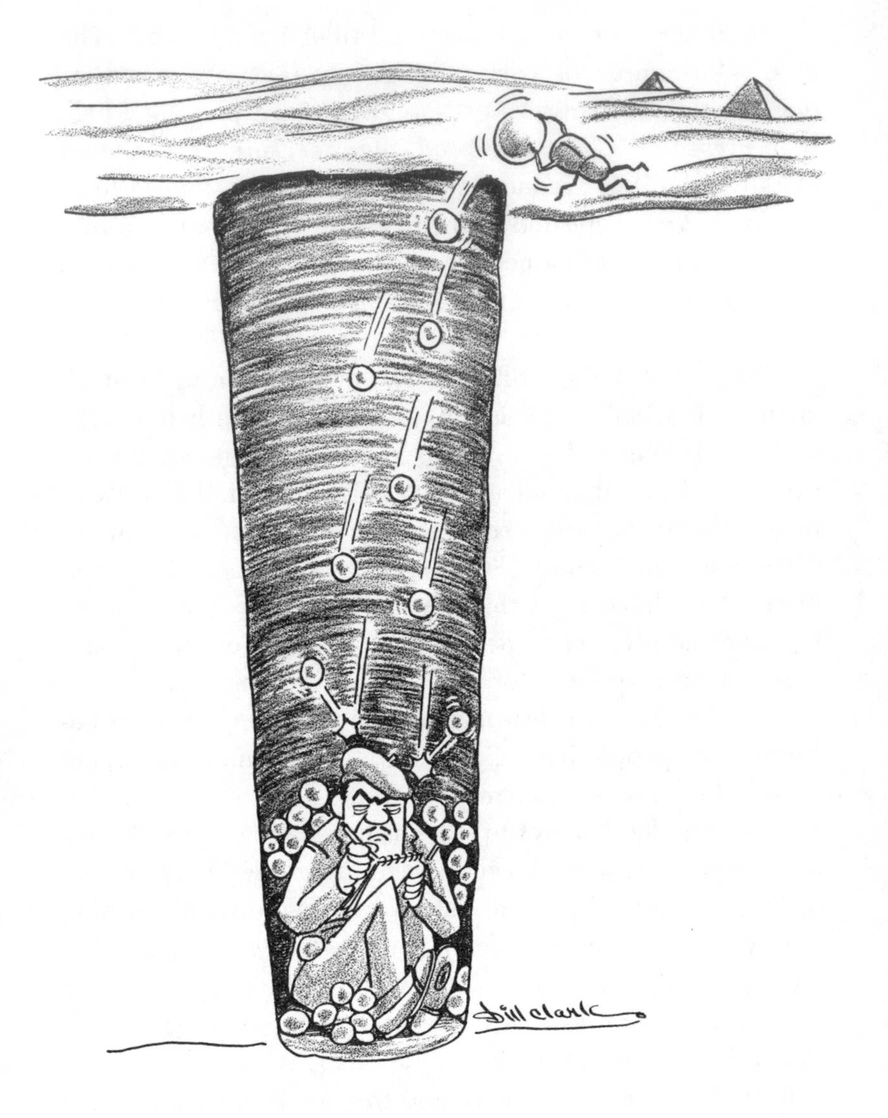

*"Deux cent soixante-six, deux cent soixante-sept, . . ."*

much more impressive performance than the student, yet it is clearly a mistake to accord the rat greater learning ability.

The fly is not the only insect that is stupid. There is a certain wasp that builds a small earthen pot astraddle a stem and provisions this pot with caterpillars which will serve as food for the wasp larva destined to develop from the eggs laid therein. If a hole is punched in the pot while it is still abuilding, the wasp will repair the hole—apparently quite an intelligent move. But if the hole is made while the wasp is provisioning the nest, she continues to poke caterpillars into the imperfect pot even though they tumble out the bottom.

There is also Fabre's classical story of the dung beetle. One species of beetles digs a shaft about five feet deep in the soil under a cake of dung. The female stations herself at the bottom. The male gathers together some dung, molds it into balls, and lowers them down to the female who tears them apart and then tramps the material into a firm substrate upon which she will eventually lay some eggs. Fabre, the villain, managed to remove the dung as fast as the female worked it. Consequently, the male kept rolling down fresh balls. The experiments ended at the 239th ball because the female died—whether from exhaustion or frustration we shall never know.

Dung beetles, incidentally, are fascinating creatures. They have inspired all sorts of people. The ancient Egyptians, believing that they arose from mud, probably saw in them a symbol of resurrection. A certain scientific society, either in jest or irony, enriches the cover of its journal with a sketch of a dung beetle rolling its ball of dung—on which is traced the map of the world.

The Čapeks in their play, "The Life of the Insects," record the following touching dialogue:

> Female Beetle: Oh, what a lovely little pile, what a treasure, what a beautiful little ball, what a precious little future.
>
> Male Beetle: It's our only joy. To think how we've saved and scraped, toiled and moiled, denied ourselves, gone without this, stinted ourselves that—
>
> Female Beetle: —and worked our legs off and drudged and plodded to get it together and—
>
> Male Beetle: —and seen it grow.

Some insects do learn, as I have already indicated in recounting their problems of navigation. Ants can be taught to run mazes, bees can be conditioned to certain colors. The famous ant of Professor Wheeler of Harvard learned to wait for the good professor to lift her up to the stage of his microscope, where there was food, instead of laboriously climbing the instrument herself. The social insects, however, are the intellectual elite of the class. Insects as a group seem to have most of their behavior built into them. It is interesting that their achievements among the invertebrate animals parallel those of man among the vertebrates. Yet they have specialized in the direction of built-in stereotyped behavior while man has become a learning animal. The chances of building a man are remote but some mechanical insects have been built that could do just about anything a bona fide insect could do except reproduce and even that may yet be possible. One of my colleagues built an electronic cockroach that rattled about the floor in perfectly good cockroach fashion.

It avoided obstacles and had convincing nervous breakdowns when confronted with insoluble problems. Under these circumstances its numerous relays would chatter angrily and eventually stop as the creature went into a mechanical sulk. In one respect it was the antithesis of a cockroach: it preferred light to dark corners.

One day in the laboratory the mechanical insect was observed pursuing a beauteous laboratory assistant down the corridor. The scientific explanation of this unforgivable behavior was that the photoelectric cells constituting the creature's eyes were stimulated by the white lab coat worn by the young lady. My friend, however, indicated in a brief remark that other explanations were possible. "I never knew till now," he explained, "what sex the creature was."

"Look on the branch above your head," said the Gnat, "and there you'll find a Snap-dragon-fly. Its body is made of plum-pudding, its wings of holly-leaves, and its head is a raisin burning in brandy."

Lewis Carroll, *Through the Looking Glass*

# Chapter 13

THE idea that the head of a fly is a raisin burning in brandy is only slightly more incredible to many people than the idea that the head contains a brain. The astonishment of people confronted with the knowledge that flies do indeed possess brains stems in large part from the fact that few people understand what a brain is, and fewer still (excepting morticians, physicians, butchers, and gourmets who feast on *cervelles de veau en matelot*) have actually seen a brain.

To see a fly's brain one needs only a reasonably powerful hand lens and an old razor blade. When the head is cut in half, the brain is revealed as a soft, apparently structureless gob ranging in color from dirty white through gray. If squashed onto a piece of paper, all of its wonder and mystery vanishes into a damp, greasy spot that quickly dries to nothing.

The brain is not in fact structureless; however, to discern its details one must carry out a long, tedious process of preparation, a process requiring great skill and artistry. The preparation of living tissues for study is one kind of scientific technology that is not mere cook-bookery. Some people can do it; others cannot.

Since the brain is soft and watery—and alive, it must be killed, and it must be rendered less watery. The killing must be accomplished quickly—not for humanitarian reasons, unimportant here because the brain has already been removed from the fly, but because slow dying results in chemical changes, which destroy the brain structure. Many chemicals suffice; the best are concoctions of formalin (a preservative for cadavers), picric acid (a high explosive), mercuric chloride (a violent poison), and alcohol.

After the brain is killed and hardened it must be cut into thin slices. Before being sliced, however, it must be either frozen or embedded in a firm matrix. As any housewife knows, cold sausage meat slices more easily than warm, and sausage in a casing more easily than unsupported meat.

*"I've overdrawn my grant budget. Would you mind slicing this fly brain for me?"*
*"Don't I need a degree to slice it that thin?"*

Today in laboratories there are slicing machines that must never fall into the hands of the baloney slicing gentry. These machines cut slices as thin as one thousandth of an inch.

If the brain has been embedded in wax, as is commonly done, the slices stick end to end as they come off the machine so that one is left holding what looks like a long thin wax tapeworm. Because of static electricity the thing is devilishly alive. Without warning it may jump from the table and wrap itself around the nearest metal object, a microscope, a door knob, or a pair of steel rimmed spectacles. It may plaster itself on one's hand where body heat melts it away to nothing. It may stick stubbornly to the table where it resists all efforts to dislodge it. Or a student, colleague, or salesman may suddenly open a door and blow the work of hours into nothingness.

Barring these mishaps, the scientist's job now is to tame this ribbon and stick pieces of it onto a glass slide. White of an egg is the adhesive commonly employed. Next the wax must be dissolved away so that the little slices of brain may be seen. A novice soon discovers to his dismay, as did also the pioneers in this endeavor, that the slices are practically invisible. To make their details visible they must be stained. The scientist has hundreds of stains at his disposal. Since different parts of tissues take up different stains, a properly stained bit of tissue rivals in color and pattern the most inspired colors in the world of nature. A designer of fabrics would find here inspiration beyond his wildest dreams.

Look at a fly brain that has been properly prepared by an expert. Its complexity, its beauty, defies description. Here are thousands of cells and nerve fibers whose arrangement

*"Do you think we flies will inherit the earth?"*
*"Who thinks?"*

makes any city road map or circuit diagram of any engineering marvel look like the constructions of a child. The poet Shapiro wrote of the fly

> O hideous little bat, the size of snot,
> With polyhedral eye and shabby clothes,
> To populate the stinking cat you walk
> The promontory of the dead man's nose,
> Climb with the fine leg of a Duncan-Phyfe
> The smoking mountains of my food
> And in a comic mood
> In mid-air take to bed a wife.

But he had never seen the brain that governs the beast. And even in the beastiality of the fly there resides a beauty because the fly cannot act otherwise. People have a choice.

Sheer numbers and interconnections render the task of understanding this brain a staggering one. It is simple for engineers to make electronic or mathematical models of simple nerve circuits or of some few behavioral acts of an animal. To espouse ultimate understanding of even so simple a brain reflects an optimism outside the natural order.

But some level of understanding can be achieved. Not only can we study brain waves of a fly, we can record multitudes of electrical changes flicking throughout these labyrinthine pathways. And as a scientific article of faith we believe that what the fly does is somehow related to these events. What Sherrington, the great English physiologist, has said of other brains describes equally well the speck, the bit of tissue, the spark that is the fly: "An enchanted loom where millions of flashing shuttles weave a dissolving pattern, always a meaningful pattern though never an abiding one . . ."

The Caterpillar and Alice looked at each other
for some time in silence: at last the
Caterpillar took the hookah out of its mouth,
and addressed her in a languid, sleepy voice.
"Who are *you*?" said the Caterpillar.

Lewis Carroll, *Alice in Wonderland*

# Chapter 14

IN a drama it is customary to present at the beginning the dramatis personae. At the beginning of this account it was not clear that it was a drama. It has had, however, its dramatic moments, although these may have lost something in being transposed from the vital, mysterious, three, or perhaps four, dimensional atmosphere of the laboratory to the two dimensional printed page. Be that as it may, the central figure, the fly, has been exposed in all his foibles. Another figure, less central but no less important has lurked in the background throughout, and though you may have become aware of him as an unseen power, the time has come to present him to you.

By the rules of the game, Science is supposed to be objective. This is, of course, ridiculous. As long as Science is conducted by scientists it will be subjective. There is a basic law of physics which states that in one way or another the observer always affects what he observes. This effect may be infinitesimally minute or beyond all hope of measurement or it may be apparent to any dullard. But clearly between the fly and the scientist there is an interaction. What the scientist does to the fly determines in part what the fly does, and what the fly does is seen by the eyes of the scientist and sends nerve messages along his optic nerves to his brain and is there switched around and juggled and changed and eventually comes out as a thoroughly subjective observation. So, perhaps to know the fly one must also know the scientist. He is the second dramatis persona, presented after the curtain falls as is the author in all proper (and improper) plays.

Strange as it may appear to the layman, the scientist is, above all, a human being. He is as inconspicuous in a crowd as a mythical Martian come to earth in human form. He has a wife and children whom he treats no better or no worse than anyone else. He experiences happiness and unhappiness like anyone else, and he gets sick and dies—like anyone else.

Outside his science he is not inevitably smarter than

everyone else. If anything, he has a modicum less horsesense than the average man on the street. In politics he may be incredibly naïve. But he probably spends a greater period of his life preparing for his career and works longer and harder at it, at a lower salary (at least in the academic world) than many of his fellow men.

Since he spends nearly as much time in his laboratory as do his own experimental animals, he tends to acquire certain amenities. The average laboratory invariably has the makings of tea or coffee. There is at least one comfortable, not infrequently ancient, chair. And some favorite art form usually decorates the walls. These objets d'art may range all the way from a sample of reverse appliqué made by the San Blas Indians to a genuine Pollock.

The central figure in this comfortable atmosphere works either in his shirt sleeves or in a lab coat. The lab coat has acquired the rank of a status symbol in the minds of the public, shiny new Ph.D.'s, and some of the older, more insecure Ph.D.'s. On the other hand, working in a laboratory is dirty and occasionally highly destructive to clothing as anyone who has had the seat of his pants burned out by acid will attest. A colleague of mine who disdained lab coats once spilled some brilliant purple dye on his shirt—one of the fast dyes. Knowing that the spots could never be removed, he stripped off his shirt with hardly a second thought and dyed the whole thing a brilliant purple. He happened that day to be wearing a green and yellow plaid tie! From then on he acquired the habit of wearing a lab coat. For the majority of scientists the coat is merely a way to protect one's clothing.

The laboratory itself is a little community whose members are, or should be, devoted to the search for truth. The number of key members in this community varies. The English and Germans do this thing well. Any good English or German laboratory has a skilled shop man, a skilled stock attendant, and a bevy of skilled laboratory assistants or technicians. The average American laboratory is fortunate to have a stock man and/or a shop man. Since in the States, for some paradoxical reason, there is never enough free money to hire qualified personnel for these glamorless but essential jobs, many labs are staffed by a motley array of singularly unskilled personages possessed with a curious affliction causing them to believe that the organization chart is an inverted pyramid. It is in the course of putting up with this sort of thing that the scientist reveals his real human qualities.

There is one laboratory of my acquaintance that once employed a particularly unendowed egomaniac as stock man. One day a long-suffering scientist in subtle revenge plucked a chicken and brought the box of feathers to the stock room saying that he wished to check in the feathers since he had no further use of them. The stock room man at first demurred but upon being assured in highly scientific language that the feathers came from a particularly rare breed of chicken he duly filed them away somewhere among the dust and trivia on his shelves. A few days later another scientist, having plucked a chicken, arrived to turn in some more feathers. The stock room man was now habituated, but his composure was shattered when he was informed that these feathers should not be mixed with the first since they rep-

*"Me? Fly specks!"*

resented an entirely different breed of chicken. And so over the course of days, different batches of feathers, allegedly of different genealogical background, and delivered by different persons were brought in to be stored. Then for a few weeks all was peaceful. The time now arrived for phase two of the operation. A scientist came to the stockroom, presented an erlenmeyer flask (the kind with the thin neck), and requested one liter of White Leghorn feathers. With clumsy fingers the stock room man slowly stuffed feathers into the flask until it was full. A few days later another member of the laboratory arrived with a request for a kilo of Rhode Island Red feathers. With a somewhat superior air the stock room attendant said, "Feathers come by the liter." "Oh, no," replied the biologist, "White Leghorn feathers come by the liter; Rhode Island Red feathers come by the kilo." The end came when someone approached the stock room with a request for one square meter of feathers. Have you ever attempted to cover a table top with a single layer of feathers while people are constantly entering and leaving the room?

The fact of the matter is that a scientist must be a jack-of-all-trades. Consider, for example, the publishing of research results. The end result of research is usually a published account of the data and conclusions. The motives for doing this are varied. For some people there is the fascination of seeing their names in print—a fascination that attains a high level of evolution in the arts. For others there is a desire, a desire that is the basis of all culture, to pass on acquired information. In any case, people love to recount their experiences. Some people feel so compulsive about this that they invent experiences. This is sometimes known as fiction. Today in science, fiction is frowned upon at all times.

The desire to recount one's experiences may be quite innocuous or it may be for personal aggrandizement. On the other hand, in some scientific circles one's promotion and permanency may depend upon the number, weight, or volume of his published works (seldom on the quality, in these institutions). A scientist I once knew was arraigned in court for threatening bodily harm to a colleague. The accused threatened to hit the defendant with his (the defendant's) published works.

For a good number of scientists, however, publication is the partial fulfillment of a desire to make available to the world a little fragment of knowledge painstakingly acquired. Since our knowledge of the world is built up of the infinite fragments of knowledge acquired by numberless individuals over all ages, the only way in which the fragments can even be pieced together into a whole, into truth, is by permitting the gleaners of the fragments some knowledge of contributions of others. For this the scientist exposes himself to the ordeal of publishing.

For ordeal it is indeed. Some people never face it so that the knowledge they gained is forever locked up in their own consciousness. To publish one must write. The muse of research is frequently a total stranger to the muse of speaking and of writing. Many a scientist is a stranger to his native tongue. In a rough sort of way there are official means of compensating for this lack. After a research worker has written five or six drafts of his message (the final one frequently bears a remarkable resemblance to the original), he selects a scientific journal to which to send the manuscript.

Ideally the journal should be selected on the basis of its subject matter and on its circulation. More often than not a given journal is selected because the lag between acceptance and publication is less than a year. Also, unfortunately, a publication is commonly selected because it has a reputation of never refusing a manuscript.

Having reached the editor's desk a manuscript then goes the round of referees. A scientist's work is judged by his peers. Reviewing manuscripts is one of the fringe duties of scientists, the majority of whom feel that with the privilege of being a scientist there are certain obligations, one of which is to give to other scientists the benefit of their knowledge in their areas of competence.

If a manuscript survives the scrutiny of referees, it receives a going-over from the editor. Editors come in all shapes, sizes, abilities, and philosophies. Some are frustrated writers with a firm conviction that they and only they are masters of the English (or German, French, etc.) language. A manuscript that has passed through their hands resembles a modernistic etching in blue. At the opposite extreme is the editor who believes that the illiteracy of the writer should be exposed for all to see. Between the two are innumerable sub-species.

A scientist is usually judged by his published works. In them he is exposing his very soul. In the long run, incompetence and fraud will out, because every statement is open to verification.

But the rewards are great. These rewards may be monetary—though this is exceedingly rare—they may be security, they may be honor. One of the more pleasant rewards is a passport to the world, a feeling of belonging to one race, a feeling that transcends political boundaries and ideologies, religions, and languages. The successful scientist has colleagues in all lands, and his work is a passport to the far corners of the world. And among his colleagues he develops lasting friendships. And with friendship comes, if not understanding, at least sympathy.

But are these the real reasons for being a scientist? I think not. The real reason I believe is more lofty and more subtle. In this I do not mean the heroics of wanting to banish pain and misery or to advance technology. Arrowsmith was really a caricature of a scientist. It is not in this manner that a scientist has nobility. It is in a far more subtle mode, a mode that many scientists themselves may never recognize, for one of the characteristics that sets man apart from all the other animals (and animal he undubitably is) is a need for knowledge for its own sake. Many animals are curious, but in them curiosity is a facet of adaptation. Man has a hunger to know. And to many a man, being endowed with the capacity to know, he has a duty to know. All knowledge, however small, however irrelevant to progress and well-being, is a part of the whole. The instrument of a scientist's destiny may be many things from the ultimate space of the farthest reaches of the universe to the ultimate particles of matter, and all things in between, not excepting man himself. It is of this the scientist partakes. A fly is just as much in the scheme of things as man. No less a person than St. Augustine remarked in the Fourth Century: "For it is inquired, what causes those members so diminutive to grow, what leads so minute a body here and there according to its natural appetite, what moves its feet in numerical order when it is running, what regulates and gives vibrations to its wings when flying? This thing whatever it is in so small a creature towers up so predominantly to one well considering, that it excels any lightning flashing upon the eyes." To know the fly is to share a bit in the sublimity of Knowledge. That is the challenge and the joy of science.